CONTROVERSIES IN ARCHAEOLOGY

Left Coast Press Inc. is committed to preserving ancient forests and natural resources. We elected to print *Controversies In Archaeology* on 30% post consumer recycled paper, processed chlorine free. As a result, for this printing, we have saved:

4 Trees (40' tall and 6-8" diameter)
1,636 Gallons of Wastewater
658 Kilowatt Hours of Electricity
180 Pounds of Solid Waste
354 Pounds of Greenhouse Gases

Left Coast Press Inc. made this paper choice because our printer, Thomson-Shore, Inc., is a member of Green Press Initiative, a nonprofit program dedicated to supporting authors, publishers, and suppliers in their efforts to reduce their use of fiber obtained from endangered forests.

For more information, visit www.greenpressinitiative.org

CONTROVERSIES IN ARCHAEOLOGY

ALICE BECK KEHOE

WALNUT CREEK, CALIFORNIA

LEFT COAST PRESS, INC.
1630 North Main Street, #400
Walnut Creek, CA 94596
http://www.LCoastPress.com

ISBN 978-1-59874-061-5 hardcover
ISBN 978-1-59874-062-2 paperback

Library of Congress Cataloguing-in-Publication Data:
Kehoe, Alice Beck, 1934–
Controversies in archaeology / Alice Beck Kehoe.
p. cm.
Includes bibliographical references and index.
ISBN 978-1-59874-061-5 (hardback : alk. paper) —
ISBN 978-1-59874-062-2 (pbk. : alk. paper)
1. Archaeology—Social aspects. 2. Archaeology—Moral and ethical aspects. 3. Archaeology—Methodology. 4. Archaeology—Case studies. 5. Prehistoric peoples. 6. Antiquities. I. Title.
CC175.K44 2008
930.1—dc22 200800354

Printed in the United States of America.

♾ The paper used in this publication meets the minimum requirements of American National Standard for Information Sciences-Permanence of Paper for Printed Library Materials, ANSI/NISO Z39.48-1992.

08 09 10 11 12 5 4 3 2 1

Contents

ILLUSTRATIONS

TABLES

PREFACE

Controversies in Archaeology is not intended to be, itself, controversial. Controversies about archaeology and controversies over archaeological finds abound. This book presents a selection of the most widely known controversies, drawing from them an explanation of scientific method and examples of critical thinking. Neither an exercise in debunking nor a promoter of pet theories, it aims to steer readers along a path between academic dogma and outré enthusiasms.

At this level of the general reader or undergraduate student, it is valuable to present something of the range of controversies encountered around a variety of archaeological topics. To go into depth on every example would exhaust the reader, so I have given more space to illustrative cases, and for others I've noted accessible sources that can be pursued by interested readers (and students doing term papers). Had I simplified the subject entirely, focusing chapters on only one or two cases and omitting an indication of the range, I would have made this "Controversies for Dummies." Learning about the much-ballyhooed famous finds should be intriguing to general readers, and students will, I hope, talk with their instructors for further details and analysis. In my own teaching, I've looked for textbooks that give a range of examples on the topics under examination, freeing me to discuss ones I find particularly significant. These may include my own research (as I use it in this book), and if the textbook offers a satisfactory range, the cases I choose to expound on can be related to the breadth of the topic. Overall, this book is designed to encourage continuing pleasure in the fascinating world of archaeology and its practitioners, while enhancing readers' discrimination between empirical data and speculation.

I thank Mitch Allen, publisher and experienced archaeologist, for inviting me to write this book, and for his careful review of the manuscript. Carol Leyba was a thoroughly pleasant and professional production editor.

Mitch and I teach undergraduates, and this is the level I write for; hence, this book does not discuss disciplinary controversies over whether X was an archaic state or a chiefdom, or what sampling strategy will yield adequate data—Professor Phuddy Duddy can write about professional debates. We hope all the US (Uncertain Students—thanks, Tim Pauketat) out there will enjoy the book, and find archaeology as intriguing as we do.

Introduction

Why should there be controversies in archaeology? Isn't archaeology a science? Surely, we would suppose, intelligent use of the scientific method will produce valid knowledge of our past. Science is like a rock held firmly to earth by gravity, but like a rock, it has a history of changes through time. Archaeology is a science remarkably suited to illustrating how societal biases and conventional dogmas affect interpretation, and even discovery, of empirical facts. This book describes and explains a good sample of controversial sites and theories, building a picture of working scientists struggling to make sense of observations.

Around us flutter all sorts of non-scientists grabbing archaeological matter to bolster ideas ranging from religion to extraterrestrials. Within the larger archaeological community are professionals more intent on career success than scholarship, entrepreneurs selling tourism, amateurs doing their best with limited knowledge—a diverse population affecting the practice and image of archaeology. Science is only part of the picture.

If applying the scientific method always gave straightforward, true answers, only nerds would do science. The fun part of science is finding unexpected data and proposing an explanation that startles people. To keep your job as a scientist, you need to make a chain of steps from data to explanation, with each step being a process that others could replicate. On the other hand, to make wads of money, you wouldn't waste your time setting out a logical chain from data to explanation. You'd just start writing, keeping your sentences short and snappy, and labeling pictures "Extraterrestrial astronaut taking off from Maya palace." Rarely, a reputable archaeologist may have the gift of writing well enough to hit the nonfiction bestseller list—Brian Fagan is one of the few.

Archaeologists walking past airport bookshops sometimes feel like nerds. That feeling evaporates when we get off the plane, go into our conference, and see our colleagues' presentations. It is exciting to talk about how pots that look Peruvian got into Mexican graves three thousand years ago. Did merchants, that long ago, sail thousands of miles along the coasts of the Pacific? Why did people in Afghanistan and Mexico and Utah make hundreds of little clay statuettes of naked women with fancy hairdos? Getting a psychic or mystery writer to fantasize about archaeological data is like whooshing down a water slide—you screech with the thrill and it's over. Doing archaeology that stands up to detailed questioning is like running—your feet must hit the ground. We can exercise our minds as we exercise our bodies, getting a runner's high from pacing through intellectual challenges.

Here in this book, we'll jog through a series of checkpoints, identifying claims made for archaeological knowledge about what earlier peoples *could* do, what they *did* do, and whether we today can actually discover what they did or thought. First, we find out that "knowing the past" is highly selective: the real past, even just the period when humans lived on earth, includes every breath of every creature, an infinitude of detail beyond actual knowing. Out of the real past, people today select bits of what's preserved and often use mute objects to justify national or ethnic claims. Meanwhile, entertainment media sell us fantasies, usually telling us it's all deep mystery. Generally unheard are the voices of the descendants of the people who made the sites and objects unearthed by archaeologists; slowly, more and more archaeologists are collaborating with these communities. Most are America's First Nations (or Australia's, or Africa's, or Oceania's); others are descendants of slaves, coolies, miners, homesteaders, or factory workers.

ARCHAEOLOGY AS SCIENCE

We move next to an examination of the scientific method and why it doesn't grind out true answers like packaged muffins. All living beings use the scientific method as we observe our surroundings, categorize information, and interpret what will happen. Even plants, as they turn toward light, are observing, categorizing, and interpreting data. Any organism that didn't sense data, relate them to categories of "good" or "bad," and react accordingly, couldn't survive. Humans put words to what we see, hear, feel, taste, and smell; make up categories that can be useless or fake ("Ten Best Slasher Movies," "Galaxies Visited by Starship Enterprise"); and sometimes interpret observations to conclude that we are doomed to hell—not a notion other animals or plants think up. The scientific method has enabled certain people to invent better techniques for producing food and protecting us from harmful cold or heat, and has enabled others to invent terrible weapons. Observing, categorizing, interpreting: the scientific method, a *process* basic to living beings.

Science with a capital "S" is a modern category, defined in the seventeenth century as procedures that are limited to actual, physically observable phenomena. This was intended to exclude what could not be observed with our human senses, or what a person claimed only he or she (and no one else) could observe. God, angels, extraterrestrial space creatures, voices from heaven, all things or phenomena believed to transcend earthly substance, are outside science, by definition. They are not proven *not* to exist, but because they're not observable in our physical world, they can't be studied scientifically. The purpose of scientific research is to understand our world more fully. If you want to talk about the ultimate meaning of life, or whether M&Ms should come in pastel colors, go ahead, but it won't be science.

Archaeology is a science. Like its entwined fellow discipline, anthropology, archaeology focuses on humans and their activities, throughout the world and through the two million years or so of humans[1] in the world. Archaeology and anthropology reach over into history, literature, religious studies, and the arts to observe human behavior and its productions, and into natural sciences such as geology, ecology, chemistry, and biology to analyze materials and understand the environments we have inhabited. The bottom line for archaeology is that we stick to observable data.

Controversies arise in archaeology over whether an interpretation is derived from observed data, and whether an interpretation is reasonable, in the light of accepted interpretations from comparable data. We have heard assertions that, for example, a large circle of stones was a sacred site: What was *observed*? Small boulders placed loosely to make a circle forty feet in diameter. Were there figurines or carvings that might represent holy beings? No. Finely made containers with residue of incense? No. Any construction that might have been used as an altar? No. Graves? No. Well, sorry, apparently there is no *evidence* comparable to items often associated with sacred places. It isn't enough to say that ancient people *could* have arranged the stones to mark off a place for ceremonies. What was recovered from the site—the *archaeological record* there—is (typically) too limited to support an interpretation of why people carried the stones and set them in a circle. Is the site, then, useless for our efforts to learn about people in the past? No, it does tell us that people were at that place, and if a highway or shopping mall or subdivision is to be built there, archaeologists will advise the developer to modify the design so as to leave the site intact, or at the least, record it as part of the regional database for interpreting the past.

---

[1] What does it mean to be "human"? Belonging to the zoological taxon (category) *Homo sapiens sapiens*? Were our earlier ancestors "human," considering they made fire and tools? One of the open challenges to archaeology is figuring out evidence that might indicate when complex language use developed, an essential feature of human societies as we know them historically.

All Manner of Controversies

Then we look at various controversies. Among the popular ones are the supposed impossibilities of raising pyramids or Easter Island statues, and fun experiments showing how feats could be accomplished with geometry and muscle. More serious controversies ask how humans first invaded Europe and then the Americas during the Ice Age, and whether warfare was involved. These issues lead us into basic controversies over human nature and the development of societies. And, of course, there are controversies over proving religious claims, from whether the Bible's tales can be linked to actual ruins, to whether archaeology disproves evolution. Controversies within archaeology include challenges to the dogma that no one crossed oceans until Christopher Columbus sailed in 1492. Readers will realize that the scavenger hunt for knowledge from the past is a quest game with rules and unexpected hazards, the joystick for navigating is your critically thinking brain, and winning the game opens opportunities for any number of new quests.

Critiquing Archaeology

In principle, we evaluate archaeological research according to how well it follows scientific method. In reality, much depends on whether the researcher has completed a graduate degree, been employed in the profession, and produced reports considered acceptable by others in the profession. One of the purposes of archaeological organizations' meetings is to assess the credibility of presenters and their projects. There is an archaeologist who habitually dressed in too-tight black T-shirts, showing off his biceps, when presenting papers at meetings; audiences could not help but wonder whether he had more brawn than brain, and quickly noticed discomfiting bits about his supposedly Greatest Oldest Site in America. Another man has been presenting surprising research on butchered mammoths in Wisconsin, dating from a time when the immense front of the continental glacier was only a few miles north; this man's quiet, serious manner, his meticulous checking of data with specialists in related fields, and the reputation he built over years of soundly conducted projects inclined listeners to accept his interpretations.

It is easier for professors in prestigious universities to win grants and get their work noticed—it's assumed that in competitive hiring, they were the best. This certainly isn't always the case, but it does mean that archaeologists in less famous universities, where they are likely to have heavier teaching and contract project loads and less institutional support, have to work harder to prove their worth to their peers. Archaeologist Jane Holden Kelley describes the archaeological profession as a set of concentric circles. The core of the set is made up of researchers who work at respected institutions or companies, who have graduate degrees, and whose projects fit within what philosopher Thomas Kuhn called "normal science," where leaders in the field have defined problems to be investigated and the core workers follow, in Kuhn's words, as "puzzle solvers." Around the core are practitioners whose research questions or methods differ from the widely shared standards, perhaps because they are innovative, perhaps because leaders of the profession consider the issue or method faulty. Kelley sees this core system as dynamic, one that shifts as innovations gain wider acceptance and some earlier commonalities fall out of favor. Individuals seen as leaders may attract followers by force of personality (their unshakable conviction that they are brilliant), although it is, thank goodness, quite possible for a solid, sensible, intelligent thinker to influence normal science.

On the fringe of a core system are fanatics obsessed with kooky ideas, a few good researchers hampered by off-putting personalities, and some excellent scholars who think outside the box, or core. Practitioners inside the core hold outsiders to higher standards than they accept for ordinary puzzle-solving work. Conversely, a thoughtful person standing on the fringe of a core system may see a parade of proclaimed emperors with remarkably few clothes, when it comes to well-grounded procedures. These extremes are part of the dynamic of real science, where, because scientists are human, anxieties and unconscious biases can intrude into clear thinking.

Ideas as well as practitioners form a core system. American archaeology's core ideas have included:

- The Americas were populated by hunters coming overland from Siberia to Alaska not long before the end of the Pleistocene Ice Age let the great glaciers melt and the land connection flood, creating the Bering Strait.
- American Indians in Anglo America (United States and Canada) did not develop civilizations, only small, usually stable societies dependent on hunting even when raising corn.
- The Americas were totally isolated from other continents after initial Bering migrations.
- Anglo America had little if any connection with Latin America.

The first core idea was reasonable, and the other three derived from propaganda downplaying the intelligence and political development of the Indian nations conquered during Anglo expansion. All four ideas have been challenged and generally discredited by subsequent archaeological research.

In the nineteenth and early twentieth centuries, American archaeology reflected racist ideas taken for granted in Anglo culture—namely, that "colored races" are inferior to Whites and were thus left behind in the evolution of civilization. This notion justifying conquest, displacement, slavery, and tearing Indian children away from their families only slowly gave way to recognition of human rights and the attainments of non-Western nations. Because archaeologists grow up and live in the larger society, and for the most part are people not engaged in political debates, they have not always been aware of the degree to which much of mainstream American archaeology, its core system, unconsciously incorporated American racism. For most of the nineteenth century, Americans generally assumed the great earth-built architecture of the Midwest's Hopewell and Mississippian mounds could not have been constructed by mere Indians. Examining this assumption was one of the first major projects of the Smithsonian Institution's Bureau of American Ethnology, and its archaeologists refuted the racist put-down through extensive surveys and test excavations. When taken-for-granted ideas are exposed as racist, good citizens (including many archaeologists) feel uncomfortable, so advancing respect for cultural diversity has been slow. Slow, but real: today, most archaeologists are open to modifying our older core ideas, and American Indians increasingly collaborate in the study of what we all acknowledge is their histories in America.

This book is about controversies. The title warns you that it *won't* tell you about the great bulk of archaeological research that is sound science. Most of the controversies we will be examining do not admit a simple, straightforward resolution. Evidence (data) may have been pulled out of context, or the function of artifacts may be debated; accuracy or significance of dates can be challenged, or possibilities of social contacts denied. Interpretations that fit popular stereotypes fare better than upsetting ones. Some controversial positions rest upon religious convictions attributed to divine revelation, by definition outside the scientific method. Reading and discussing controversies in archaeology is training in critical thinking, yet it will also raise consciousness of the power of social forces upon individuals' thinking. Through such nuanced assessments we gain competence in dealing with our real, complicated world. Although archaeology is about the past, it's carried on in the present, and it can sharpen our skills for daily and future life.

A Note about the Author

The first thing you, the critical-thinking reader, need to know about this book is that its author is, one, a reputable experienced archaeologist, and two, a person not quite routinely accepted into American archaeology's core system. Because I obtained my Ph.D. the year the Civil Rights Act passed the U.S. Congress, I did not have many of the opportunities presented to my male peers. Generally, hiring and salaries openly favored men, and at least one of my graduate school professors of archaeology told us he would not accept women doctoral candidates in his program. Since I had attended an outstanding women's college, Barnard, I knew (from empirical observation) that women could be as good scholars as men. Experiencing unjust bias firsthand heightened my thinking critically about the status quo, whether about the people acclaimed as leaders (like the professor who would not accept women) or about core ideas. It happened, too, that while at Barnard I worked as a student aide in the anthropology department at the American Museum of Natural History, where I observed a fine archaeologist and scholar, Gordon Ekholm, rebuffed when he put forward evidence for pre-Columbian trans-Pacific voyages. My undergraduate exposure to the scrupulous science of Ekholm, his colleagues James Ford and Junius Bird, and my professors Richard and Nathalie Woodbury helped me distinguish between sound work, sloppy thinking, and fads in archaeology.

Somebody said I am "edgy." I hope he meant "on the cutting edge of research." In this book, I present controversies on which most archaeologists agree with my evaluation, and others about which fewer would fully agree. The point is not to persuade you, the reader, that I'm the all-seeing guru, but rather to lay out the principal points and logic of controversy. Follow out the cases through further reading, judicious use of the Internet, and if possible, visiting sites and museums to see the real thing.

REFERENCE

Kelley, Jane H., and Marsha P. Hanen, 1988. *Archaeology and the Methodology of Science*. Albuquerque: University of New Mexico Press.

NOTE: Sources listed for the chapters contain references to earlier publications. Students should use the listed sources as jumping-off places for further research. For a general introduction to archaeology, I suggest the inexpensive *Archaeology: A Concise Introduction*, 2007, by Alice Beck Kehoe and Thomas C. Pleger (Long Grove, IL: Waveland Press), and for information on North American Indian prehistory and ethnohistory, including Mexico, I recommend my own *North American Indians: A Comprehensive Account*, third edition, 2006 (Upper Saddle River, NJ: Prentice-Hall). For information on sites, time periods, and artifacts, an encyclopedic source is *The Human Past: World Prehistory and the Development of Human Societies*, 2005, edited by Chris Scarre (London: Thames and Hudson).

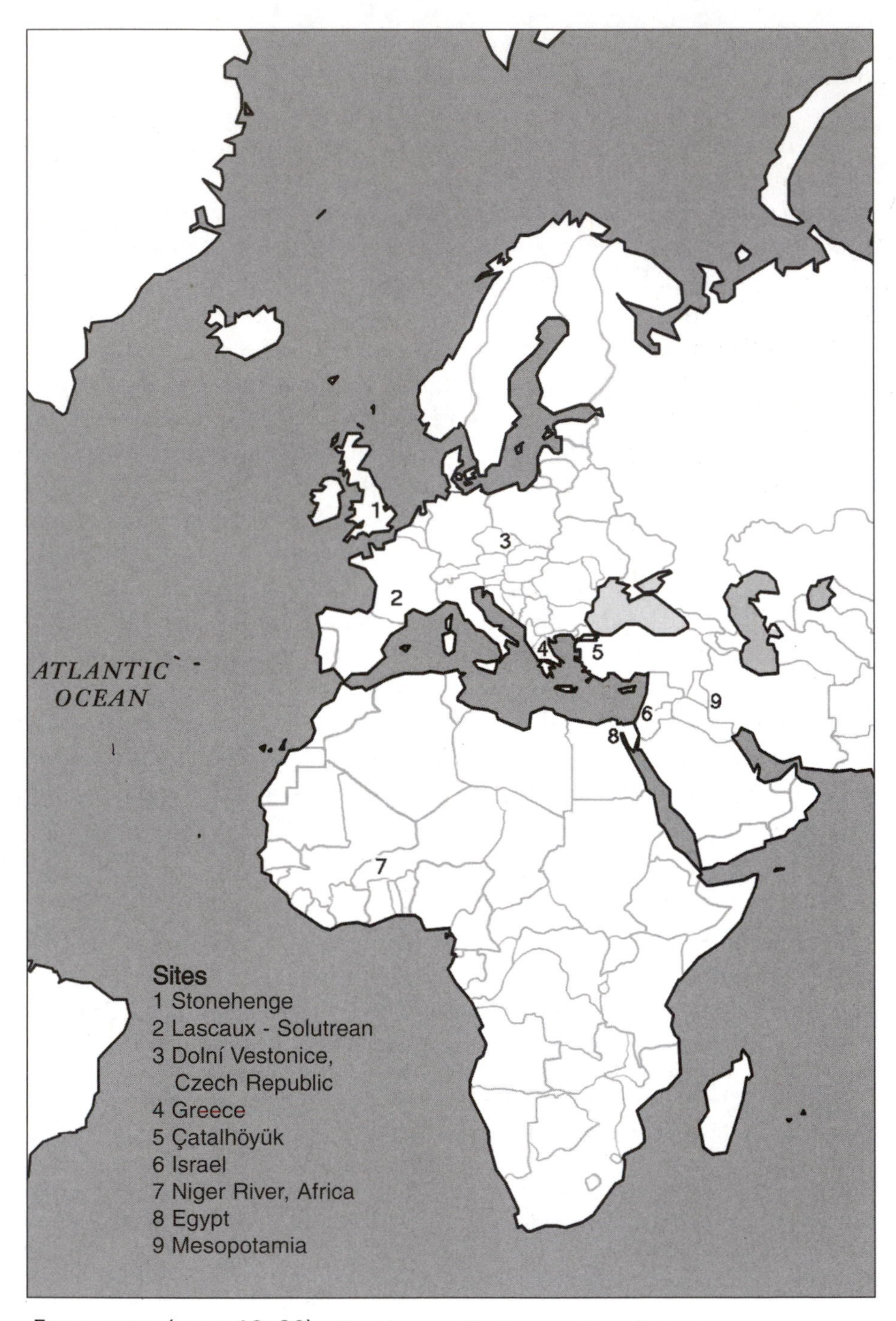

FRONTISPIECE (PAGES 18–20). Global map with sites mentioned in text.

ARCTIC OCEAN
12
PACIFIC OCEAN
11
10
Sites
10 Thailand
11 China
12 Siberia, Bering Strait

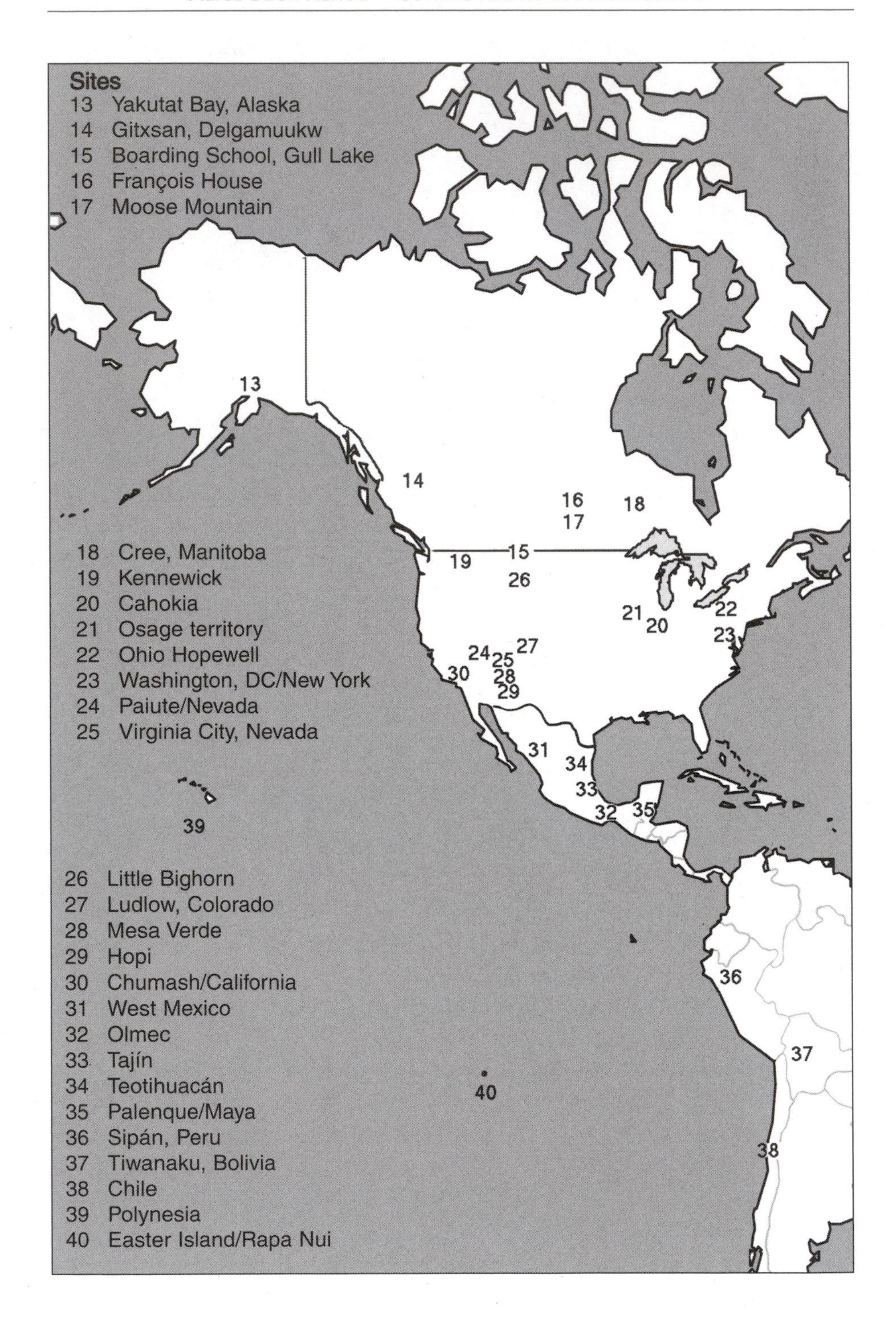
Sites
13 Yakutat Bay, Alaska
14 Gitxsan, Delgamuukw
15 Boarding School, Gull Lake
16 François House
17 Moose Mountain
18 Cree, Manitoba
19 Kennewick
20 Cahokia
21 Osage territory
22 Ohio Hopewell
23 Washington, DC/New York
24 Paiute/Nevada
25 Virginia City, Nevada
26 Little Bighorn
27 Ludlow, Colorado
28 Mesa Verde
29 Hopi
30 Chumash/California
31 West Mexico
32 Olmec
33 Tajín
34 Teotihuacán
35 Palenque/Maya
36 Sipán, Peru
37 Tiwanaku, Bolivia
38 Chile
39 Polynesia
40 Easter Island/Rapa Nui

1 The Past is Today

What does it mean to be human? One part of being human is to know we have a past. Other animals have memories without, so far as we can understand them, the mental concept of "the past," a whole world that once existed but no longer exists as it was. Literate societies have, for centuries, recorded the world in writing and visual images, other societies in orally presented history and tales and in images. Relics from the past have been collected, wondered at, or put to use. Europeans thought stone ax heads found in fields might be thunderbolts. The Aztecs in Mexico brought offerings of ancient figurines and ornaments to the Great Temple in their capital city, Tenochtitlán. Other vestiges of the past have been so familiar they were overlooked: prehistoric earth mounds unnoticed because they seemed just part of the landscape, scatters of potsherds ignored like recent broken pottery. Systematic searching for remains from the past is part of modern science.

British archaeologist Glyn Daniel remarked that "the past" can be construed in different ways, and he warned against naively accepting a "wished-for past," a picture of a past age that is more daydream than history. Robin Collingwood, a professor of philosophy at Oxford who was also a practicing field archaeologist, taught that the *real* past cannot be known: the *real* past had uncountable gazillions of animals, plants, bacteria, viruses, grains of sand, molecules, raindrops, snowflakes, zaps of lightning—you get the picture. "Known pasts" are put together out of material relics, transmitted knowledge, and assumptions; the interplay between data (always, of course, far less than once existed) and stereotypes produces a series of projected "known pasts." We have privileged written documents from the past over other artifacts as evidence for the past, creating (in the nineteenth century) the profession of historian to glean a known past out of these texts. That left most of human existence and huge areas of the world as "peoples without history," a disrespect archaeologists challenge. From stains in soil, undecayed remnants of artifacts, food, and shelter, landscape modifications, and biological effects of habitat and behavior seen in skeletons, archaeologists construct pictures of settlements, and by looking at style distributions in time and space, estimate political organizations. With written texts, as in Mesopotamia from 3000 B.C.E. or by the Maya in Mesoamerica, we can put names to gods and rulers; without texts, the people in the past are nameless but no less vividly known to archaeologists.

Our predecessors recognized "pasts" and the fact that changes occurred in the past. Classical Greeks described a Golden Age when living was easy and luxurious, from which societies had degenerated into corruption. Mesoamericans and Hindus preached a series of earthly ages, each demolished by cataclysms, followed by resurrections created by deities. Christians took the Old Testament literally: patriarchs living nine hundred years, the Tower of Babel, Noah's flood, and all (see Chapter 6). They did not make these claims without some evidence. Greeks occasionally found gold ornaments from Mycenaean times a thousand years previously; Mexicans picked up strange figurines and ancient temple remnants they matched to local memories of destructive eruptions and floods; and Christians claimed various ruined walls and eroded valleys resulted from biblical events. A favorite theme in Western culture is a contrast between our alienated materialistic society and a wished-for past of contentment with simple pleasures, a Garden of Eden. This could be melded with a wished-for geography where Noble Savages live happily, with simple pleasures and deep true spirituality—Classical Greeks proposed they were Scythian nomadic pastoralists on the steppes of Asia and eastern Europe, and some Europeans viewed American Indians as mythical Noble Savages. The myth supposes that Noble Savages never change—that is, they are "the living Past," evidence of a more general way of life in past time, a way of life led by the ancestors of our own alienated, materialistic city-dwellers. American Indians have to put up with well-meaning tourists visiting contemporary reservations wanting to see living-fossil Stone Age Man and maybe discover True Spirituality, only to come away disappointed that Indian people live in tract houses, watch TV, and eat potato chips. The past is gone.

"Known pasts" live in the present. This chapter looks at the issue of "who owns the past." Is it the descendants of the people whose remains are excavated, or the modern government administering the territory? Or perhaps all humanity? This latter view led the United Nations to designate a number of places as "World Heritage Sites." Known pasts are used to draw tourists, often making up a substantial part of a local economy. Tourists may want to see "the real thing," provoking disputes over whether antiquities ought to be subject to pollution or feet pounding on a landscape ought to be permitted. Some communities feel tourists may pay more money to see re-creations in a theme park. Other uses of the known past involve private collections of antiquities; commercial dealers sell attractive objects, often for many thousands of dollars, and wealthy persons may buy antiquities as investments, figuring they are as good as gold in being readily convertible to cash if needed. But this high-end market for ancient art encourages looters to vandalize archaeological sites in search of marketable objects, never mind the destruction of irreplaceable scientific data.

■ ■ ■ NATIONALISM

"Our past" has been essential to modern nationalism, which narrows the term "nation" to a physical territory with a distinct language. Modern nations assert their inalienable right to their homeland by exhibiting archaeological finds from the territory, proving that people did live there for millennia. Right—but that doesn't really legitimate the nation: Were the ancient people the forebears of the present population? Sampling DNA from populations has revealed much more interbreeding than fervent nationalists want to admit. Does it matter?

National museums function as shrines. They select the most beautiful objects from the past in their territory, some of which may actually be imports (but if so, demonstrate the good taste and buying power of imputed ancestors). They may have maps with arrows showing migrations or timelines marked with new dynasties, implying how desirable it always was to come live in their nation. Architectural displays and preservation districts complement museum messages, generally downplaying international architectural styles in favor of the local.[1] Museums may display artifacts from other cultures to contrast them as crude against their own society's art heritage, as anthropologist Stephanie Moser discovered the British Museum had done in the eighteenth century with Egyptian antiquities. English eyes looked at this art as primitive next to Classical Greek and Roman sculpture though it often predated classical objects by thousands of years.

---

[1] See book by Silverman for "art" versus "ethnic" in museums, and by Moser for British attitude toward Egyptian antiquities.

Scientific archaeology has its own nationalistic roots. A method for systematically and carefully discovering and recording vestiges of the past was launched early in the nineteenth century in Denmark, after that country had been disastrously defeated by Napoleon's navy. King and people rallied around the idea that, although they had just lost a good part of their former territory in Sweden, they had a substantial past no one could take from them. Royal patronage supported a new National Museum that was assiduously filled with antiquities dug from Danish soil, arranged by the curator to show progression from a primitive Stone Age through an Age of Bronze (the first use of metal) to an Iron Age and the present. This three-age framework demonstrating Progress attracted interest throughout Europe, making Denmark admired just as the king had hoped. Excavators could see the sequence in the ground: iron and steel tools just below the surface, bronze tools lower down, and stone tools lowest of all. The sequence also fit Enlightenment (the eighteenth-century intellectual movement) logic, having most ancient humans using the simplest tools of natural materials, followed by people manipulating easily worked copper and alloying it to make bronze, and eventually metallurgists figuring how to smelt and work iron for cheaper, better tools. In the 1840s, about the time Danish-style archaeology was being taken up throughout the rest of Europe, a French avocational scientist, Jacques Boucher de Perthes, presented papers describing stone tools embedded deep in gravel banks along the Somme River Valley, along with fossil bones of mammoth and other extinct beasts. Boucher de Perthes assumed the animals and their hunters had all perished in Noah's Flood. After Charles Darwin published his *On the Origin of Species* in 1859, a biblical explanation lost favor. A delegation of British scientists visited Boucher de Perthes, examined the Somme gravel banks with him, and declared their opinion that his stone tools were artifacts from the Stone Age and thousands, maybe millions, of years old. Human culture evolved as did animals and plants, from ancestors through descent with modification. Natural selection applied to us as well as the rest of the natural world.

While scientific archaeology and its sister sciences geology and paleontology were developing field methods and classifications, antiquities from Mediterranean and Western Asian cities were retrieved from palace and temple ruins and brought to Europe for display. Much of the region was within the Ottoman Turk empire, successor to Byzantium, with its capital at Istanbul (Byzantine Constantinople) where the Black Sea outlet runs into the Mediterranean. Invading from central Asian steppes in the fourteenth century, Turks overran ancient Mesopotamia (southwestern Asia), Egypt and adjacent North Africa, Greece, and southeastern Europe. Their method of government was to permit local religions, languages, and legal customs to continue, so long as people respected the Ottoman governors' decrees and paid taxes. Such relatively tolerant rule let nationalistic sentiments persist. As Europe shifted out of medieval feudal states into larger, centralized modern nations, Turks controlling the eastern Mediterranean were a major threat. Along with their political, military, and commercial power, they controlled the cities and art treasures of Classical Greece, Egypt, and Mesopotamia—the foundations of Western civilization, revered since Europe's fifteenth-century Renaissance. European intellectuals wanted to rescue these antiquities from the infidel Turks.

In 1799, Thomas Bruce, seventh Earl of Elgin, was named British Ambassador to the Ottoman court. He used his post to save—he claimed—the beautiful marble sculptures adorning the Parthenon Temple in Athens. The building itself, more than two thousand years old, was damaged in 1687 when a bombardment by attacking Venetians hit gunpowder stored in the temple by Ottoman defenders. More than a century afterward, Lord Elgin felt justified in stripping off the carved panels and statues and shipping them to England "for safekeeping." He offered them to the British Museum to improve public taste, a noble gesture except that he asked over a hundred thousand dollars for the treasures. After several years of haggling, the Museum bought "the Elgin Marbles," and they have adorned the London museum ever since (Figure 1.1).

Meanwhile, Greeks fought the Ottomans, achieving independence in 1829. Their rebellion engaged romantic European intellectuals, including English poet George Gordon, Lord Byron, killed by fever in a rebel Greek camp in 1824. Wars of independence by southeastern European territories against the Ottoman Empire fueled nineteenth-century Europeans' picture of Asian states as "backward," despotic, ignorant of science, and mired in mystifying spiritualism. Now labeled "orientalism," myths about Asians colored interpretations of Asian (and Egyptian) antiquities and histories at the same time that they bolstered Europeans' adoration of Classical Greek and Roman relics. They conveniently disregarded the fact that medieval Arab scholars in Baghdad, Cairo, and Damascus preserved the Classical texts of the Greeks and Romans and conducted substantial scientific research while Europe was mired in the Dark Ages. Nor did Europeans recognize the contributions that Chinese science and technology contributed to Europe—gunpowder, watermills, magnetic compass, and paper, to name a few.

As tourism grew into one of the globe's largest industries after World War II, Classical Athens's acropolis drew millions of visitors, becoming a significant part of Greece's contemporary economy. Tourists expected to see the famous sculptures of the Parthenon. Greece asked Britain to return the Elgin Marbles.

FIGURE 1.1 Parthenon temple sculpture, Athens, Greece, ca. 440 B.C.E. *From George Redford*, A Manual of Ancient Sculpture, *London 1886*.

The British Museum replied "No," claiming its charter forbids giving away its collections. Greece appealed on grounds that Lord Elgin never had *Greece's* permission to remove the art; furthermore, the Ottoman permit Elgin did obtain is rather ambiguously worded, perhaps meant to allow only removal of loose fallen stone. Greece's most glamorous actress, Melina Mercouri, passionately spoke for the return of the Marbles. Nothing availed; although Germany did return a small piece of the frieze in its possession, more than half of the frieze of horsemen, fifteen statues, and over a dozen other Parthenon sculptures remain as the centerpiece of the British Museum.

Where do the Elgin Marbles belong? In the hands of the British, who claim to have legally obtained them and have preserved them for the past two hundred years? Or do they belong to the Greeks, who claim them as their heritage and want them back in their original context? Clearly, this is a controversy between competing national interests.[2]

[2] See book by Brodie, Kersel, Luke, and Tubb, and by Gholi Majd, for more on antiquities trade cases and the practice of awarding excavation permits to European powers.

WHO SHOULD EXCAVATE SITES? WHO HAS THE RIGHT TO KEEP ANTIQUITIES?

Archaeology worldwide was first practiced by a few western European and North American nations. It has been customary for excavation projects in less-developed countries to be financed by wealthy patrons and institutions from major world powers, with recovered artifacts divided between the host country and the excavators. Supposedly, each gets half of the "best" as well as of ordinary material. Ever since these arrangements began, in the nineteenth century, host countries have complained that they get the commonplace stuff and foreigners the "goodies." France, Germany, Britain, the United States, and other Western nations formally arranged "concessions" whereby they had exclusive privilege to excavate important sites, or a region, or even all archaeology in a country. For example, in 1900 France was granted a monopoly on archaeological work in Persia (present-day Iran), with the right to export expedition finds without customs inspection. This lasted until 1927, when pressure from the U.S. government on France resulted in a new agreement in which a Frenchman remained as Persia's Director of Antiquities, but American museums were allowed to carry out excavations. An American government representative wrote in 1935 that the Persian antiquities transported in great quantities to U.S. museums are *America's* "national treasure . . . contributing to the cultural advancement of the American people" (quoted in Gholi Majd 2003:187). Ironically, the University of Chicago Oriental Institute insisted, after the 2003 outbreak of war in Iraq and of Near Eastern hostilities implicating Iran, that Chicago is really the proper custodian for its collections from the region, because it can preserve them from destruction in their embattled homelands.

Controversial as are "custodial" arrangements that never end, loans for exhibits that don't come back, and divisions of finds that don't seem fair, the picture darkens more when we consider illegal digs and black-market dealings.[3] There are many families in antiquities-rich areas whose income derives from looting graves and ruins and selling the attractive objects illicitly to dealers (see below, "Who Benefits from Archaeology?"). Some of the objects may be bought by museums from dealers claiming they had been exported before passage of laws protecting antiquities (see box, "Antiquities Laws and Trade"). Most of the objects go to private collectors, in some cases solely as investments likely to

[3] In an article on the antiquities trade, Kathryn Walker Tubb (2006:285) quotes the dust jacket of a 1995 book by Peter Mould about the market for "lost" fine paintings: the "high-risk, high-stakes game of art dealing—a game of hair-trigger intuition, of poker nerves and pumping adrenaline, of fierce rivalry and ruthless competition."

Antiquities Laws and Trade

Most United Nations member states signed the 1970 UNESCO Convention on the Means of Prohibiting and Preventing the Illicit Import, Export and Transfer of Ownership of Cultural Property, and the 1972 Convention Concerning the Protection of the World Cultural and Natural Heritage. Most nations, including the United States, have laws regulating ownership of their antiquities, making prohibited action a criminal offense. Since 2002, U.S. Sentencing Guidelines recommend heavier penalties, both in imprisonment and in fines, than had been used previously for persons looting or vandalizing sites and/or dealing in the black market in antiquities. This has created a difference between objects in the art market prior to 1970, which are considered legal antiquities, and those looted from archaeological sites after 1970.

In 2001, well-known New York antiquities dealer Frederick Schultz—president of the national association of dealers—was indicted for conspiracy to steal and sell antiquities from Egypt. Schultz schemed with an English expert in restoring art objects to smuggle objects out of Egypt and to counterfeit documents to show that the objects had been bought by a (nonexistent) wealthy collector decades before, trying to circumvent the international conventions established in the 1970s. The conspirators even covered with gaudy paint the actual head of Pharaoh Amenhotep III, from 1351 B.C.E., to make it look like a tourist souvenir, and then tried to sell it for $1,200,000! Schultz was convicted, sentenced to nearly three years in prison, fined $50,000, and forfeited the illegally gained antiquities. The Englishman was tried in Britain under that country's laws and also convicted.

Schultz's conviction came under the U.S. emphasis on antiquities as property, in his case property owned by Egypt and stolen from its owner, the Egyptian state. Fraud and forgery in trade of course entered into the case, establishing that the conspirators knew that what they were doing was illegal. British law has a different emphasis, that of the priceless nature of antiquities, how they are part of a country's patrimony. The Schultz case, only one of many, exposed the huge profits an unscrupulous dealer or collector could realize, an issue that the United States considers significant, while Britain considers the harm done to the origin nation's patriotic knowledge and symbols. Most recently, the emphasis has moved toward major Western museums who were complicit in this trade, by not examining too closely the origins of ancient objects they purchased on the art market. The Metropolitan Museum of Art, Boston Museum of Fine Arts, and Getty Museum have all returned objects to Italy and Greece in recent years after being confronted with evidence that the objects were looted.

increase in value. This angle makes it ever more profitable to fake antiquities. Archaeologists need to know the *provenience* of finds—exactly where found, and how obtained—to understand their context, essential to interpreting age, ethnic affiliation, and ways of life. Looted objects lose context, and thus their value for science and history, and encourage the proliferation of fakes on the art market. Until recently West Mexico's past had little professional archaeological investigation and was known primarily through unprovenienced, large clay figurines popular with art collectors, objects that may well be mostly fakes (Figure 1.2). A well-publicized major museum exhibit of Olmec figurines may have been one-third fakes; when a knowledgeable archaeologist asked the curator about them, that person explained that if the museum refused to show these famous pieces, often pictured in art books, their wealthy owners would be miffed and might cease to donate to the institution.

FIGURE 1.2 Figurine of a leader, Colima, West Mexico, ca. 100 B.C.E.–A.D. 250. Clay, 19.5 inches high. *Amerindian Art Purchase Fund 1997-363. © The Art Institute of Chicago.*

Objects claimed to be antiquities lose considerable monetary value if they turn out to be recently made, no matter how artistic the piece; some might be appreciated as art, but would sell for much less if not purportedly ancient. Authenticity for archaeological finds requires documentation of the artifact *in situ*—that is, in its original site—preferably documentation by the discovering archaeologist using photographs as well as fieldnotes and field drawings and maps. Such documentation gives context, too, which allows the archaeologist to understand the object in light of the lives of the people who originally made and used it. Authentication of museum pieces sold to the institution or to a donor by a dealer or local middleman depends upon specialists such as art historians, who judge whether the style, material, and technique conform to provenienced pieces. If all the unprovenienced or poorly documented objects in museums and private collections were thrown out, thousands of real antiquities from older purchases would be lost along with the confusing hoaxes; museums would be denuded, left with not much more than drawer upon drawer of mundane potsherds and stone tools from professional scientific excavations.

Like the art specialist, a skilled faker will closely study these objects, reproducing them and "antiquing" the creation with acid, short-term burial, or other ingenious manipulations. Thus, we have ongoing contests between honest dealers and collectors, on one side, and people hoping to earn good money through astute fakery. A down-on-his-luck Georgia man who had spent countless hours mastering the art of flint-knapping offered a dozen beautiful stone spear-points in the ancient Clovis style to a well-known collector-dealer in Santa Fe, telling him they had been found by a deceased local man years before. The dealer called in two prominent archaeologists specializing in studying Paleoindian stone weapon points; impressed by the perfect, difficult Clovis-style technique, the experts judged the points authentic. (See Chapter 8 for more on Paleoindians.) Still, the dealer was super cautious because over a hundred thousand dollars were at risk. He sent the points to another Paleoindian researcher who could conduct chemical tests. A wise move: suspicious traces of modern Teflon, differences between red ocher and what was only red dirt, and lack of minute weathering expected on eleven-thousand-year-old artifacts prompted the dealer to phone the Georgia man. Fearing criminal conviction for attempted fraud, the flint-knapper apologized and returned the dealer's money.

Collecting antiquities as pieces of art is like going to see the Parthenon because it's listed in a tourist guide, without finding out why it's famous. The Elgin Marbles are more than superb sculptures; they illustrate an impressive civic ritual held in a lively, ambitious city two thousand years ago, a city nurturing philosophers, scientists, orators, artists, technicians, and military strategists whose works are still part of Western culture. "Heritage tourism" is a growing segment of the global tourist industry, recognized under the World Trade Organization's Global Code of Ethics for Tourism. Controversies arise over a number of issues: whether what the tourist sees is authentic; how to define "authentic"; how to interpret the site for tourists; protecting sites from damage; protecting tourists from damage; ensuring that local communities gain from the business; and ensuring they have some say in the enterprise. At one extreme will be visitors who are more content to hang with bored surfer dudes in abbreviated Roman costumes at Caesar's Palace in Las Vegas than to traipse around the real Coliseum in Rome. At the other extreme are people backpacking along Inca roads high in the Andes. In between are hundreds of millions of vacationers watching their wallets, wanting clean comfortable beds and not-too-exotic food, eager to view architecture, sites, and objects they learned about in school or on television's Discovery and History Channels. What they *don't* want is controversy. Yet, the acquisition policies of major museums, the dollars paid for ancient art, and the robust forgery industry make heritage tourism just that.

When the Rockefellers backed restoring Colonial-period buildings in Williamsburg, Virginia, it was taken for granted that they would finance public buildings, including a tavern, and homes of respected citizens. Years later, after the Civil Rights Act of 1964 encouraged less powerful groups to speak out, Colonial Williamsburg extended its work to include slave quarters, and hired African-American historians and interpretive guides offering tours showing the lives of free and enslaved Blacks in Williamsburg and a nearby plantation. Most visitors respected that this side of the colony should be told, although the majority continued to prefer the standard "Whites" tours, but a few felt their hero-worship of Revolutionary War burghers was affronted by revealing the underprivileged lives in the back quarters. The situation has been worse at Thomas Jefferson's mansion, Monticello. Excavation and partial restoration of the slave quarters artfully hidden from the mansion (slaves went through a tunnel to their work in the house) reminds some visitors of the authenticated gossip that Jefferson fathered children by one of his slaves, and *kept those children in slavery* at Monticello. Guides at Monticello have been advised not to bring up this shocking flaw in the Founding Father. Interpretive centers commemorating Lewis and Clark's 1804–06 epic trek across the West similarly do not highlight the children fathered by the Corps of Discovery with Indian women along the route. Their descendants on the reservations can tell you about them, but local heritage tourism overlooks that legacy. How much should the past be sanitized for the tourist? If history is inherently contentious, where do communities draw the line in portraying these controversies?

■ ■ ■ REGULATING TOURISM

Access to archaeological sites is another source of contention. The most famous case is Stonehenge, the four-thousand-year-old site in England of huge upright slabs of stone set in a circle.[4] So many thousands of tourists walked around and inside the circle that the stability of the stones was threatened. It became difficult to see the circle for the crowds. The British government finally prohibited walking in or close to the monument, setting up approved walkways and cable fences. Now visitors are disappointed that they must keep a distance, and those who come on account of alleged mystical forces or to worship as they suppose their very distant ancestors did, get quite angry. On Midsummer's Eve (summer solstice), police are prominent around the site to quell celebrants, those inspired by religion and those inspired by alcohol. Part of the excitement comes from the popularity of Stonehenge knock-offs—carhenges of upended half-sunk automobiles, half-size stonehenges adorning campuses, cartoons galore—and dramatic photographs, one being a desktop background image packaged with Microsoft software. Stonehenge is pop culture as well as popular tourism, so limiting access to it is as annoying as it would be to limit access to other British icons—say, recordings of the Beatles or the Rolling Stones (no relation to the prehistoric Standing Stones).

Because of their symbolic value to nations, religions, and ethnic groups, mere access to some sites can spark conflagration. When a prime minister of Israel insisted on visiting a mosque in Jerusalem on a spot believed by both Muslims and Jews to be holy, fragile agreements between Israel and Palestinians collapsed in a storm of violence. Hindus and Muslims in India have unleashed bloody clashes at historic temples used by one or the other but claimed by both. American Indians chose in 1973 to stage a protest against injustices at the site at Wounded Knee, South Dakota, where in 1890 a village of Lakota had been massacred by U.S. troops. United States marshals were dispatched in 1973 to put down what the government called a rebellion, maintaining a siege around the hamlet for weeks, shooting and being shot at, and killing two protesters. History that makes a site significant may still inflame passions, and passions, in

---

[4] See Chippindale's *Stonehenge Complete* for everything known and speculated about England's most famous ancient site.

turn, excite ethnic group oppositions and government efforts at control. Archaeological controversies have been deadly.

The four-hundredth anniversary of the English settlement at Jamestown, Virginia, 1607, was touted as the "Founding of America." Millions of dollars were invested in excavating the original fort and also the nearby Indian town, headquarters of the Powhatan kingdom in whose territory the English settled. Virginia Indian communities were enlisted to advise on interpreting and reconstructing Powhatan's town, and to play their forebears in a film. A charming young girl portrayed Pocahontas and reenacted the tale of her embracing Captain John Smith as her father the Powhatan ordered him killed (a story told years later in Smith's memoirs, impossible to verify). Some of the Virginia Indian communities are disturbed that the Englishmen's landing is "celebrated," because for them, Jamestown was an illegal takeover of Powhatan's land, and the colonists' murders of dozens of Indian people when, during a drought year, they refused to sell their corn, marks the beginning of genocide. These Indian descendants say Jamestown's beginning can be "commemorated," but not "celebrated." Although this point of view is acknowledged by the Jamestown development enterprise, tourist income from the "celebration" is counted on to repay the state's investment in "America's Founding Colony," though technically, that title belongs to St. Augustine in Florida, a quarter-century older.

■ ■ ■ WHO BENEFITS FROM ARCHAEOLOGY?

Professional archaeologists like to think of themselves as good-hearted people helping poor foreign communities by hiring laborers for excavation. Villagers may beg the archaeologist to be godparent for a child or honored guest at a wedding. It's not always without ulterior motives: such ritual roles obligate the archaeologist to give gifts, and may entail years of requests to assist in educating children, buying a pickup or tractor, or recommending villagers for jobs. Sometimes an archaeologist feels exploited.

What do villagers think? Here comes the archaeological team, highly educated foreigners or upper-class citizens of their own nation; now, for three months, they pay wages, good by local standards, to a select number of village men to labor and a few village women to cook, clean artifacts, and perhaps babysit; then they are gone, gone are the wages, gone are the artifacts taken from the villagers' ground. In the long term, the archaeologists gained "capital" (study material) to invest in furthering their careers in their prosperous cities, whereas the villagers gained little and lost the capital represented by the archaeological material so long buried in their territory.

One long-standing response by villagers has been to use their knowledge, honed by experience with the archaeologists, to excavate on their own, not scientifically but to find objects to sell. We are horrified. They get sullen. It's reported that communities in North Coast Peru believe their ancestors left their remains for their descendants' economic needs; from this point of view, archaeologists taking finds aren't much better than sixteenth-century Spanish conquistadors taking their gold (Castillo and Holmquist 2006:155). Tomb-robbing to make a living goes back to ancient Egypt—those fabled curses over pharaohs' tombs were meant to deter the robbers. Today's *huaqueros* (Peruvian term for illegal diggers) may have little else to lift them from dire poverty. And although they get a pittance for what they sell, compared with the millions the big-time dealers reap, it does put food on the table and repair the roof.

More and more, archaeologists are actively concerned with giving back to their field sites. One American archaeologist working in Mexico years ago paid her workers wages comparable to a laborer's wage in the United States—much more than the local day wage. Several of the men were able to save enough to buy sewing machines with which their families produced clothing for the market, a cash income supplementing their subsistence farming. In another case, in the 1980s, an American working in Bolivia, near Lake Titicaca in the Andes, interpreted odd ridges in lakeside lowlands to have been ancient raised fields. He persuaded villagers to rebuild several ridges to their original height (by shoveling dirt on top from the ditches between ridges) and gave them seed potatoes to plant on them, to test his hypothesis on the function of the ridges. His hypothesis was confirmed; the villagers got bumper crops, and they and their neighbors rebuilt more ridges (one widow built her own little ridge in her yard). Within a few years, the government of Bolivia was teaching raised-field agriculture to hundreds of rural communities, not always situated suitably for the technique. A similar experiment with similar results was carried out at that time on the Peruvian side of Titicaca by another American archaeologist.

Beginning around 1990, archaeological projects have become accustomed to include plans to leave educational exhibits on their work in local communities, in schools or civic halls, or in purpose-built buildings that serve as local museums. Because these little museums usually include exhibits on local flora and fauna too, they are often termed "eco-museums." In some instances, the archaeologist and community work with ecologists to develop nature preserves or restore degraded environments. Communities near highways hope to attract tourists to see the museum for a fee and purchase local crafts; between inspiration from seeing ancestors' ceramics and carvings, and the prospect of selling to tourists, a number of villages have revived disappearing crafts, even reinventing technology and styles forgotten in recent centuries.

Democracy is tunneling into some Third World peasant villages to the degree that they dare to demand the national government build fine local museums to display spectacular finds, not take them to the national museum. On the North Coast of Peru, huaqueros in 1987 discovered richly furnished tombs near the village of Sipán. Gossipers leaked the news. Peruvian archaeologist Walter Alva was called in, and after a tense confrontation with the huaqueros, he began salvaging the disturbed portions of the site. Soon *National Geographic* latched onto the gold-bedecked "Lord of Sipán," tourists came, a site museum was built to explain Moche culture to them, and North Coast Peruvian archaeology boomed as foreign and Peruvian archaeologists investigated more Moche sites. By 2002, the Peruvian government constructed an urban-quality Royal Tombs of Sipán Museum (Figure 1.3), with a Mochica artisans' workshop and sales space beside it, not in the village but in the regional capital of Lambayeque. This benefited the city, but without much trickling down to Sipán village.

Meanwhile, in the highlands east of Sipán, villagers in Kuntur Wasi successfully insisted on retaining in the community artifacts excavated from their locality's "royal tombs." At the suggestion of the archaeologists, a community Cultural Association was organized to negotiate the arrangements to allow the fabulous objects, comparable to the Lord of Sipán's treasure, to travel to Japan for exhibit; in return, interested Japanese raised thousands of dollars to build a sturdy museum and archaeological workshop in the village. The Peruvian government agreed to bend patrimony laws to permit all the Kuntur Wasi artifacts, even the gold crowns, to remain in the custody of the village organization, securely housed in the museum they operate. This is one community that does not tolerate huaqueros. At Kuntur Wasi, the riches left by ancient ancestors are *invested*, not sold, by their descendants, to maintain a cooperative enterprise for tourism and support a spirit of pride and legitimate ambition.

■ ■ ■ ARCHAEOLOGY CAN'T AVOID POLITICS

Political repercussions from controversies over archaeological sites or objects demonstrate that the past is part of today. It might seem that controversies over whether work is properly scientific could be calmly decided, but it is perilous to draw a line between the social dimensions of archaeological work and the scientific methodological dimensions. Careers hang in the balance, and as this chapter tells you, so may national pride and lucrative tourism. Wealthy collectors

FIGURE 1.3 Tomb of the Lord of Sipán, Peru, ca. A.D. 300. *Photo by Alice B. Kehoe.*

may disdain archaeologists' pickiness in wanting to find documentation for objects that multimillionaires bought for their eye-appeal. Emotions can be triggered even where one would expect rational discussion, at a meeting of archaeologists. At one national conference, an archaeologist cornered me and raised both fists, bellowing that I had insulted him by pointing out it is unscientific, and basically racist, to lump all non-Western religious practitioners under the label "shaman." How dare I call him a racist!

Political, economic, social, ethnic, and national interests undergird our understanding of the past. How do archaeologists weave between all these interests in the search for knowledge of our predecessors? In the next chapter, we will lay out considerations for evaluating archaeology as a scientific pursuit.

SOURCES

Brodie, Neil, Morag M. Kersel, Christina Luke, and Kathryn Walker Tubb, editors, 2006. *Archaeology, Cultural Heritage, and the Antiquities Trade.* Gainesville: University Press of Florida.

Castillo Butters, Luis Jaime, and Ulla Sarela Holmquist Pachas, 2006. Modular Site Museums and Sustainable Community Development at San José de Moro, Peru. In *Archaeological Site Museums in Latin America*, edited by Helaine Silverman, pp. 130–155. Gainesville: University Press of Florida.

Chippindale, Christopher, 1994. *Stonehenge Complete.* 2nd edition. London: Thames and Hudson.

Gerstenblith, Patty, 2006. Recent Developments in the Legal Protection of Cultural Heritage. In *Archaeology, Cultural Heritage, and the Antiquities Trade*, edited by N. Brodie, M. M. Kersel, C. Luke, and K. W. Tubb, pp. 68–92. Gainesville: University Press of Florida.

Gholi Majd, Mohammad, 2003. *The Great American Plunder of Persia's Antiquities, 1925–1941.* Lanham, MD: University Press of America.

Moser, Stephanie, 2006. *Wondrous Curiosities: Ancient Egypt at the British Museum.* Chicago: University of Chicago Press.

Preston, Douglas, 1999. Woody's Dream. *The New Yorker* 75(34):80–87.

Silverman, Helaine, editor, 2006. *Archaeological Site Museums in Latin America.* Gainesville: University Press of Florida.

Tubb, Kathryn Walker, 2006. Artifacts and Emotion. In *Archaeology, Cultural Heritage, and the Antiquities Trade*, edited by N. Brodie, M. M. Kersel, C. Luke, and K. W. Tubb, pp. 284–302. Gainesville: University Press of Florida.

World Trade Organization Global Code of Ethics for Tourism: http://www.world-tourism.org/code_ethics/eng.html

2 Scientific Method

Archaeology relies on scientific methods to produce valid interpretations of our pasts.

Here lies another can of archaeological-controversy worms. What *is* "scientific method"? Are there different scientific methods for different fields of science? Is archaeology *its own* field of science, or an area of research that calls upon various sciences? How certain can we be about events in the past, when what we study is long dead and usually left us no written record? How are archaeological interpretations different from those proposed by promoters of wild theories about the past?

The Domain of Science and the Scientific Method

Science deals with the things of this world. Science is *empirical*, testing ideas against reality. That which comes within the domain of science must be *observable*; or to put it another way, in doing science, people are interacting with the material world. Anything that has no physical existence—for example, "luck"—is outside the domain of science. Note that "thinking" can be scientifically investigated because it has material existence in electrical activities within a brain.

Just as we humans can never totally know the immense real past, we can never simply observe. We receive many more sense impressions than we can consciously recognize. While you're reading this, you're not noticing your heartbeat, the accustomed feel of your clothing—sense data we don't need to pay attention to. You probably can't say what roofing material is used on your building, information you can neglect because someone else is responsible for maintaining it.

We observe what has significance for ourselves. Young scientists are likely to consciously observe phenomena that their professors instructed them to consider significant, and to construct experiments or, in archaeology, choose sites or artifacts[1] to study that seem likely to reward observation with insight into interesting questions, frequently questions that prominent researchers formulate.

Being oriented toward consciously observing certain kinds of data, if only because the totality of sense impressions would overwhelm us, scientists say that their work is inevitably *theory-laden.* Not only do scientists use research questions to choose what observations they will note and analyze, they are also affected by their culture's values. Archaeologist Denise Schmandt-Besserat wondered about all the little clay balls in a Mesopotamian site she was working at; her professor told her they were very common in ancient Mesopotamia but no one paid attention to the plain little objects. Schmandt-Besserat figured that they wouldn't have been made if they were nothings (her theory: numerous objects of a recognizable type must have had a function). She observed and classified hundreds stored in museum collection drawers, noted a few marks repeated over and over on the balls, looked inside cracked balls and found tokens, noted the find contexts were merchants' locales, and interpreted the balls to have been records of merchants' shipments: sealed inside each ball was a token of the kind and amount shipped, such as "ten sheep." The transport carrier delivered the sealed ball to the destination merchant, who cracked it and checked the token against the delivery. In this way, writing developed in Mesopotamia over five thousand years ago, to symbolize products shipped to other cities. With agreed-upon symbols signifying kinds and quantities of goods, merchants could communicate information about shipments without having to travel with them; in time, the idea of communicating through symbols pressed into clay was extended to warehouse inventories and eventually to making symbols stand for speech sounds rather than names of things. A long-standing research question, "How did writing develop?" oriented Schmandt-Besserat toward noticing the clay balls had marks resembling a few of the cuneiform symbols later used in Mesopotamian clay-tablet writing. Her observations were theory-laden in that they followed from her hypothesis that the marks were precursors of developed Mesopotamian cuneiform writing.

---

[1] Experiments in archaeology are limited to questions of technology or short-term effects, since we don't live long enough to set up and observe processes that take generations to resolve. To compensate, we can search for "natural experiments," where similar conditions have occurred in more than one place or culture but are far enough apart that the societies involved are unlikely to have directly influenced one another.

To take a more famous example, Charles Darwin collected and classified a great amount of data, guided by his hypothesis (theory) that variations in animal and plants are results of natural selection. His theory has proved enormously fruitful, leading to the sciences of genetics and ecology as well as stimulating biology.

The obligation to observe can be misapplied. It is not scientific to arbitrarily neglect some of the data in favor of those that fit a supposed universal law or popular causal factor. There are controversies in archaeology that result from some writers' allegiance to a narrow explanation, refusing to seriously consider nonconforming data and explanatory hypotheses. Historian David Quinn remarked of a renowned Harvard professor, that he was an

> outstanding historian whose great merit is that he sees sharply in black-and-white terms and is therefore uniquely qualified to expound what is already known. He is perhaps too impatient to study the nuances. (Quinn 1974: 22–23)

Inference to the best explanation doesn't give us eternal truths; it does keep us empirical and properly modest, as befits fallible human creatures.

■ ■ ■ More Principles of Scientific Reasoning

People tend to say that science deals with things that are observable and *measurable*. This follows from science observing that which has material existence: if something physically exists, it can be measured. Counting and measuring are procedures within science but are secondary to the basic criterion that science deals with what is observable.

Scientists usually *classify* (or "categorize"). But, just because a person classifies doesn't mean he or she is doing science—one can classify angels or kinds of luck, or dragons or unicorns. More importantly, human languages all classify observations in the very way that words frequently refer to classes of phenomena. You know the word "book" means a class of objects with pages bound in a cover. Human languages go beyond scientific procedure when words are used metaphorically—for example, when the computer I am using is called a Powerbook. It's the size of a book and has content between covers (these very words I am typing and you are reading), but it isn't the same thing as a bound book of pages. Colloquial (everyday, ordinary) language is rich and evocative because it classifies observations into fuzzy and overlapping sets. Science, by contrast, strives to construct clearly demarcated sets.

Science requires *precision*. Not only in measurements, but also in words and terminology. Ordinary language gets by without precision. As I'm typing this, my cat Felicia is on my computer. Is she pressing the key to write 666? (She's done that, the mark of the beast, a couple of times!) Or is she *sitting on* my Powerbook? Yep, that's what I'm observing—you were onto me! Scientific communication is dull to read because it must carefully exclude colloquialisms with wider or fuzzier reference than the phenomenon described. It is the scientist's obligation to not only make careful observations and classifications, but also to take great care in choosing words, phrases, and essay style. Each scientific discipline has its style of presentation, some using mathematics and others, including archaeology, preferring words.

Scientists frequently employ graphs and tables to communicate concisely and take advantage of the power of visual images. Archaeologists use photographs, maps, and drawings of soil layers, layout of artifacts and features in an occupation layer, and drawings of artifacts highlighting technological detail. This technique of presenting visual images is termed *virtual witnessing*, showing data to the reader as if the reader were in the field or laboratory, standing beside the scientist. Virtual witnessing is the crux of science.

■ ■ ■ INTERPRETATION

Inference to the best explanation is the goal of scientific archaeology. Cool graduate students call it IBE. First, IBE emphasizes *inference*, "bringing in" information from observed phenomena (*in ferre* is Latin for bringing in). *Inference begins with empirical observation*. It does not begin with postulating "if *x* causes *y*, then I should find that cases of *y* are preceded by *x*"; that is, inference does not begin with a hypothesis. Instead, *inference produces a hypothesis*. The researcher observes situations, notes certain recurrent or surprising features, and proposes an explanation. This tentative explanation is an explanatory hypothesis which then can be tested with new observations (data) to see whether they consistently associate with the hypothesized cause or necessary related factor.

This is somewhat different from laboratory-based sciences, where the hypothesis often drives the research. A hypothesis is proposed, conditions are kept uniform (often in a sterile test tube), and one variable is changed for each experiment in a carefully controlled environment. Archaeological sites don't fit neatly into test tubes.

Second, IBE warns us that we seek the *best* explanation, not The Truth, nor simple validity (more than one explanation may be valid as a logical argument). IBE recognizes that we humans are fallible, liable to be misled by limited information or our emotions. More information may let us see that a once-reasonable explanation is no longer tenable because it can't account for the new data, or relies on premises now known to be erroneous. An example is the explanation that human populations in the Americas descended from Asians migrating over Beringia, the land now drowned under the Bering Strait, and moving south between the Rockies and the edge of the continental glacier covering northeastern America. Geological and paleo-climate research revealed that there was no such ice-free corridor in the late Pleistocene; Paleoindians moved *northward* from the American Plains into the supposed corridor as glacier ice melted. An alternate explanation hypothesizing migrants traveling along the Pacific coast until they reached unglaciated southern land and could spread inland, became the better explanation (see Chapter 8 for more on this controversy).

We must strive to obtain as much information as we can, and to examine our inclinations to be pleased or repelled by an explanation. Does a certain explanation seem good because it's something we always knew? It's been conventional to identify more-or-less triangular pointed stone tools as "projectile points": since we were little kids we've seen countless pictures of American Indians and Ice Age humans holding spears and arrows with stone tips. Now you're an archaeologist and you find several dozen stone "points" near a hearth in what appears, from signs of a wooden framework and tamped floor, to have been a prehistoric home. Given the *domestic* context of the stone "points," it is better to *infer* that they were kitchen knife blades, a *working hypothesis* you then *test* by examining the edges of the blades under magnification. You may then *confirm* your hypothesis with observations of working edges chipped as is typical on blades made and used experimentally for food processing; or you might *disconfirm* your hypothesis by observing impact fractures typical of points experimentally projected on spears or arrows. Whichever explanation fits your observations, you have worked scientifically instead of just assuming popular stereotypes are true. The *best* explanation develops from—is inferred from—what seem to be significant or surprising features of field or laboratory material and then testing a reasonable explanation against additional data.

It was fashionable during the 1960s and 1970s to invert scientific reasoning by starting with a hypothesis and then going out to look for data to validate, or confirm, the hypothesis, much like the lab scientist. This was termed the *hypothetico-deductive method*, or H-D, *deducing from*[2] the hypothesis what data to seek. Popular among archaeologists for two decades under the label "Processual Archaeology," the underpinnings of H-D didn't hold up. H-D proponents didn't seem to worry about how they made a hypothesis to begin with.

[2] *Deduction* means taking ideas *from* a statement or hypothesis (*deduce* comes from the Latin for "leading from"). *Induction* (Latin for "leading into") means taking ideas from observations, leading *to* a hypothesis. Inferences are made by induction from observations.

They would make up a logical proposition—for example, that people in the North American Southwest stopped building round pithouses and started to live in above-ground rectangular adobe houses because the climate changed. Then they went out with paleo-climate researchers and found evidence for climate change—voila! they had "validated" their hypothesis. But was it the best explanation? They didn't discuss whether it was, because their hypothesis only specified "climate change."

If they had tested alternative working hypotheses, they would have noted that round pithouses built a couple feet deep in the earth are much better insulated than rectangular above-ground adobe houses; the round sunken houses are warmer in winter and cooler in summer. Whichever way the climate changed, the time-tested traditional pithouse would suit. A more recent, and better, explanation begins with observations of rectangular above-ground adobe houses brought into the Southwest through contact, or possibly population movements, from Mexico; the houses are associated with maize agriculture as an economic staple and with town life. Inferring from these data in nearby regions that ancestral Pueblos acculturated to this vibrant, alternative lifestyle finds confirmation in the Pueblos' shift to reliance on maize agriculture even during unfavorable climate change. H-D can deliver neat logical cause-and-effect explanations because it doesn't encourage simultaneous testing of alternative working hypotheses. It doesn't confront the researcher with a range of potentially relevant data. A fundamental requirement for good science is that researchers consider all data bearing on their question—in this case, evidence for trade with the towns of above-ground adobe houses—and show why they think the alternative explanations are less likely.

■ ■ ■ EVALUATING INTERPRETATIONS

Archaeology begins in the field with uncovering remains of past living. Researchers move into the laboratory with everything they can take from an excavation: artifacts, soil samples, animal and plant remains, photos, field maps, drawings, remote-sensing data records, and notes. The work of analyzing all these data can be shared, and usually does bring in specialists from related fields of geology, sedimentology (soils), zoology, botany, and chemistry. Data as exposed in excavation or in laboratory examination are, as the word "data" (Latin plural for "given") indicates, given to the analyst. The analyst labels what is seen—that is, names it; classifies it by putting it in with other data judged similar, or separated because judged to be different; and suggests how the observed data once functioned in the lives of people who once lived at the site or used the artifact. Naming what is observed is the first step in a *chain of signification* leading through classification to interpretation, placing the data in a model or paradigm—in plain words, relating the data to images known from history or living societies. Hazards along the procedure include putting a wrong name on an artifact or feature—for example, calling a bone weaving shuttle a "spearpoint"; putting data into wrong categories, and calling up a model or paradigm that doesn't fit—let us say, describing the site as a kill where a man threw a spear, when careful consideration of the weight, shape, and polish of the bone artifact indicates its find spot denoted a woman weaving. Using critical thinking, the researcher figures out chains of signification and evaluates whether they contain unwarranted jumps.

Probability is another criterion for evaluating interpretations. Something that happens frequently is more likely to happen than something that rarely occurs, so an explanation involving frequent behavior is more probable than one invoking an extraordinary event. Still, "improbable" is not "impossible." It was improbable that a poorly educated, lower-middle-class Austrian would get elected der Führer (The Leader) of Germany, start a world war, and instigate the murder of millions of handicapped and ethnic human beings, but it did happen when Adolf Schickengruber moved to Munich and changed his name to Hitler. Anything that's physically possible is scientifically possible, whatever the odds. The archaeologist, as a scientist, considers the odds when deciding the best explanation, and when new data change the odds, reconsiders. One professor tells students to imagine a thermometer registering the degree, low or high, of probability for an explanation. Or you can imagine a balance scale, tipping one way or the other as data for, and data against, an explanation are piled on. At the same time, the balance scale isn't a dumpster. An early European scientist, William of Ockham (in England, thirteenth century A.D.), warned against stringing too many possibilities into a hypothesis. "Don't multiply items in a proposed explanation!" His advice came to be called "Ockham's Razor": the simplest, most straightforward explanation is generally best.

Scientists and Their Audiences

A fascinating study by historians Steven Shapin and Simon Schaffer (1985) recounts the battle, in late seventeenth-century London, between middle-class philosopher Thomas Hobbes and wealthy nobleman Robert Boyle, a founder of the Royal Society of London. Boyle supported a laboratory with technicians in his mansion, inviting other gentlemen to witness experiments performed by his staff under his direction. The audience could then confirm the success of the experiments (when they succeeded; if they didn't, faulty apparatus or incompetent staff could be blamed). Hobbes, a commoner, was not admitted to Boyle's mansion. Instead of actually witnessing an experiment, he had to accept Boyle's published description and the invited gentlemen's assurance that Boyle's paper did accurately describe the experiment and its outcome. Shapin and Schaffer expose the problem with the notion that "the public" can judge scientific claims. Boyle could restrict "the public" evaluating his experiments, excluding a man he feared would be unfriendly; today, scientific experiments are usually made public through highly technical reports available only in laboratories and university libraries. The general public doesn't see the experiments or the reports, and if people did, most couldn't figure out the technical terminology.

Robert Boyle was influential in establishing modern science. He insisted that science deal only with what could be presented to, and observed by, witnesses; he ruthlessly discriminated between observable "matters of fact" and "hypotheses" that could not be directly tested before witnesses. In his day, untestable, therefore *metaphysical* ("more than physical"), hypotheses included whether the particles making up physical matter were tiny indivisible blocks, or infinitely divisible; during following centuries, inventions in microscopy and atomic physics made the question testable. Science is not a fixed domain but one that widens as researchers figure out ways to observe postulated phenomena.

Boyle's announced practice of using credible witnesses (that is, not just his paid staff) to verify his experimental claims remains a critical part of science. Several controversies in archaeology have been resolved by the researcher inviting notable colleagues to a site to personally witness the excavation features. After such visits, L'Anse aux Meadows, dating approximately A.D. 1000 in Newfoundland, was accepted as the first known Norse colony settlement in North America; Monte Verde in Chile was confirmed to be earlier than Clovis Paleoindian sites of 12,000 years ago; and most famously, the existence of humans in the Americas by the end of the Pleistocene (Ice Age), ten thousand years ago, was acknowledged by the 1927 visit of scholars to Folsom, New Mexico, to see excavated Pleistocene bison bones with a stone spearpoint between the ribs. Most scientific claims are less hotly debated, and it suffices that researchers publish a detailed description of the evidence and the argument for their interpretation. The most controversial claims tend to be ones challenging basic "received wisdom"—for example, that Norse stories about attempted colonies in America were only myths, or that Indians arrived in the Americas after the end of the Ice Age, less than ten thousand years ago.

Historian of science Thomas Kuhn, in his landmark 1962 book *The Structure of Scientific Revolutions*, used the word "paradigm" for standard "received knowledge" in a scientific field, and wrote that substantial challenges seem to turn the world upside down—Europeans (Norse) in America before Columbus! Indians coming into an America still occupied by immense continental glaciers and herds of mammoths! Mind-boggling. Then some people, most effectively a visiting group of respected and open-minded scientists, together view and discuss the data, and publicly conclude that we must accept a new "paradigm." We could say that there's a new game in town.

Boyle's reliance on confirmation by a group of credible witnesses continues by means of *peer review*: scientific reports are read by knowledgeable specialists before they can be accepted for publication, and researchers are expected to present ongoing work at scientific meetings where they can answer questions and challenges from their colleagues. Through this method, a consensus develops between groups of scholars about interpretations of the past, based upon common perceptions of reported data. A paradigm is born.

Even peer review can't work perfectly in an imperfect society. Ideally, scientists are unselfish seekers after truth. In our real world, some are; some are honest but fearful of being criticized; and some are egotists driven by ambition. Peers reviewing one of my papers objected to my charge that standard terminology calling American First Nations "bands" or, if more populous, "chiefdoms," is unscientific because it comes out of academic convention, not well-grounded ethnographies (see Chapter 10). My point in writing the paper was that these terms connoting "primitive" social organization adversely affect legal decisions in First Nations' claims, as was evidenced in the notorious 1991 Delgamuukw case in British Columbia, Canada. There, several Indian communities lost timber rights around their villages, their principal economic resource, when a judge ruled that their forebears had been "primitive nomads" who couldn't have "owned" territory. My peers who reviewed the paper were discomfited by my claim for substantial political and economic consequences for what most archaeologists like to think is non-political science. Editors who rejected the original version relied on peer review by archaeologists not familiar with the Delgamuukw case, while the editor of the *Anthropology Newsletter* where it was finally published, and who is a political anthropologist, recognized the validity of my argument.

Peer review is the best method we can devise to deliver useful critiques to researchers and to develop criteria for judging potential value, but it also tends to enforce conformity to majority opinion, occasionally stifling really innovative work. Many controversies in archaeology boil down to conflicts between peer reviewers and researchers convinced that their peers do not comprehend their data's significance. Some journals publishing research papers in the natural and physical sciences have been using the Internet to post submitted papers and reviewers' comments on the journal website, inviting readers to add comments. In this way, papers rejected for publication are still available for reading, and instances of incompetent or self-serving reviewers may be detected. Posting data sets online is another twenty-first-century practice that broadens virtual witnessing and the audience who can comment. There's more opportunity to argue for overturning "received knowledge" paradigms, and in most countries today, no one need fear being arrested for challenging orthodoxy, as was Thomas Kuhn's example, Galileo, in seventeenth-century Italy.

The use of the Internet has its down side too. Not just scholars, but anyone can post their findings, or simply their interpretations, without regard for the filter of questions and concerns of other specialists. Atlantis underneath Tampa, Florida. The continent Mu off the coast of Japan. Noah's Ark sitting on a Caucasian mountain. A dream that locates the resting place of the lost Ark of the Covenant. Anyone can post a claim, no matter how outrageous. Peer review, then, becomes a key factor in evaluating the quality of claims about the past. Crackpot theorists don't submit to peer review. And even if innovative work is sometimes discouraged by reviewers, ultimately the long-term accumulation of reviewed evidence is likely to win out.

An Example of Scientific Method: The Gull Lake Bison Drive

To illustrate scientific method in archaeology, I'll describe a project from "the good ol' days" of the twentieth century, one recent enough to include radiocarbon dating and contemporary emphasis on thorough analysis of all artifacts and organic remains (and one that I know a good deal about). This project aimed to establish a chronology for artifacts in the northwestern plains of North America, and to explore techniques of bison hunting and processing. The project director was Thomas F. Kehoe, assisted by me, your author and his wife.

Throughout the Holocene (the modern climate era of the last ten thousand years, following the Pleistocene Ice Age), human subsistence on the northwestern plains—the Canadian prairies and adjacent North Dakota and Montana—relied on bison[3] herds. At first recorded contact with Europeans, at the end of

[3] "Bison" is the correct term for American "buffalo." They are popularly called "buffalo" because European explorers thought they looked like African buffalo. Bison are closely related to cattle.

the seventeenth century, Indians' principal method of hunting bison was to build a corral (pound) under a bluff or at the end of a ravine, and drive herds into the trap where they could be shot at close range with arrows (Figure 2.1). Our research question became: How old is this technique of hunting?

Thomas Kehoe pursued this question first on the Blackfeet Reservation[4] in north-central Montana, where he could interview elderly Blackfoot on what their grandparents had told them about this traditional method of hunting (Figure 2.2). He obtained grants to hire a trained field assistant and a crew to excavate for two summer seasons at a location on the reservation where butchered bison bones and stone artifacts indicated a slaughter site. This wasn't one of the best-known legendary bison pounds, but as a smaller site, and convenient to town, it made a more practical research site. After two seasons' excavation and laboratory

FIGURE 2.1 Bison driven into corral, Boarding School site, Montana. *Drawn by Jerry Livingston for* Plains Anthropologist Memoir *4.*

[4] *Blackfoot* is the name used for the Blackfoot alliance and for the Canadian Blackfoot groups. The Montana reservation for the southern group is called *Blackfeet* Reservation.

FIGURE 2.2 Blackfoot Indian elders (*top*, Theodore Last Star) examining Boarding School bison drive excavation. *Photo by Thomas F. Kehoe.*

analysis, we had a stratigraphic sequence of layers of butchered bison bone, stone arrowpoints, knives, and choppers going back to around A.D. 1200 by radiocarbon dating, with sterile soil strata (no bones or artifacts) in between. We knew, from laboratory examination of butcher cuts on bones, how the people had processed their kills. We even found some of the wooden corral poles at the edge of the slaughtering area, preserved by being rapidly buried under dirt washing down the bluff after frantic bison dug hoofs into the hillside trying to stop their fall. Bringing Blackfoot elders to see the excavated layers, and reading European explorers' accounts, helped us interpret what we found.

After completing this project, Kehoe moved to Saskatchewan, the middle of the three Canadian northern plains provinces. His new job as provincial archaeologist was to survey the province for sites, work with avocational archaeologists, and initiate research projects to inform the province's citizens of their heritage. One question was whether the sequence of artifact styles and bison processing discovered five hundred miles southwest in Montana was characteristic of Saskatchewan plains prehistory. For this question, he selected a site near the town of Gull Lake in southwestern Saskatchewan; avocational archaeologists had been told of this probable bison pound by the homesteading farmer, had tested it, and recommended it as clearly stratified. I was field director for the project's first season (Tom had to travel around the province quite a bit, to check reports of sites and on a couple of surveys).

We could see, from the surface, that bison had been caught on a terrace under a bluff. A semicircle of lush grass, standing out from the rest of the pasture, indicated where a corral had stood, its decaying wooden fence and refuse from butchering enriching the soil within the corral semicircle. First, we made a contour map of the terrace and bluff, including a permanent benchmark. Next, we used a surveyor's transit to lay out, and mark on the map, a grid of little wooden stakes north-south and east-west, five feet apart (later, we American archaeologists switched measuring units to the international metric system). Excavation would take place within these marked squares, so that all finds and soil layers could be measured in three dimensions from corner stakes and recorded exactly on the site map. Our crew used sharpened shovels to strip the grassy sod from the top of a line of marked-out squares, and then began carefully scraping off soil using archaeologists' beloved standard tool, sharpened masons' trowels. Soon, hundreds of butchered bison bones were exposed (Figure 2.3), along with sharp little stone arrowpoints and knife blades and heavy cobblestone choppers. We photographed each square layer by layer, drew bones and artifacts to scale on graph paper, recorded exact find spots in our notebooks and on the site map, and identified soil types by reference to Munsell soil charts. The final step was lifting the bones and artifacts into bags marked with grid square number and stratigraphic layer number; diagnostic artifacts, such as points and blades with base style intact, and potsherds, each got their own bag marked with precise find spot.

In the laboratory during the winter, we examined the bones to figure out the butchering procedure used. In bison pound sites, practically all the bone is bison, other than a rabbit or two or mouse caught in the stampede, so we quickly proceeded from identifying species to analyzing the cultural information about how kills were processed. It was clear from the clustering of heaps of butchered bison, layer after layer under the bluff, that the historic technique of driving herds into constructed pounds was used time and again, and as the Gull Lake squares got deeper and deeper, ultimately down to twenty feet below surface, radiocarbon-dated material from the series of occupation layers showed that pounds had been made here for two thousand years. Laying out excavated arrowpoints according to the layer in which they occurred informed us that

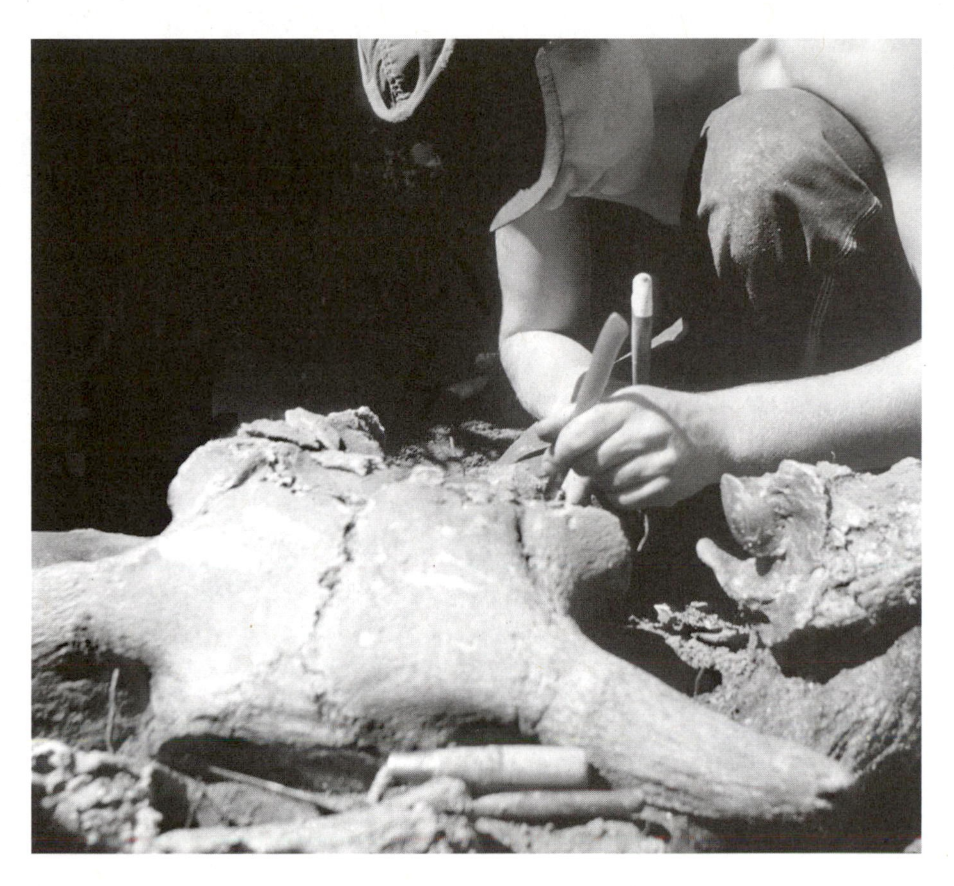

FIGURE 2.3 Excavating bison skull from the Gull Lake bison drive. *Photo by Thomas F. Kehoe.*

styles of arrowpoint had changed over the centuries, from a type called Avonlea (after another Saskatchewan site where the style was first identified), through Prairie Side-notched styles to the Plains Side-notched style used historically (Figure 2.4). These styles had been identified by several archaeologists, including Tom, comparing sequences of arrowpoints from historic upper layers down through buried earlier layers in a number of northern plains sites. Using this sequence of styles, random finds in fields or eroded banks can be dated.

The very lowest layers at Gull Lake (above glacial till of which the hill is formed) were thin bands of charcoal probably from campfires, with few artifacts or bone. These layers do not show evidence of pounds. Did the people making Avonlea-style points introduce the technique of driving herds into pounds, nearly two thousand years ago? Evidence *for* this hypothesis is that the distinctive

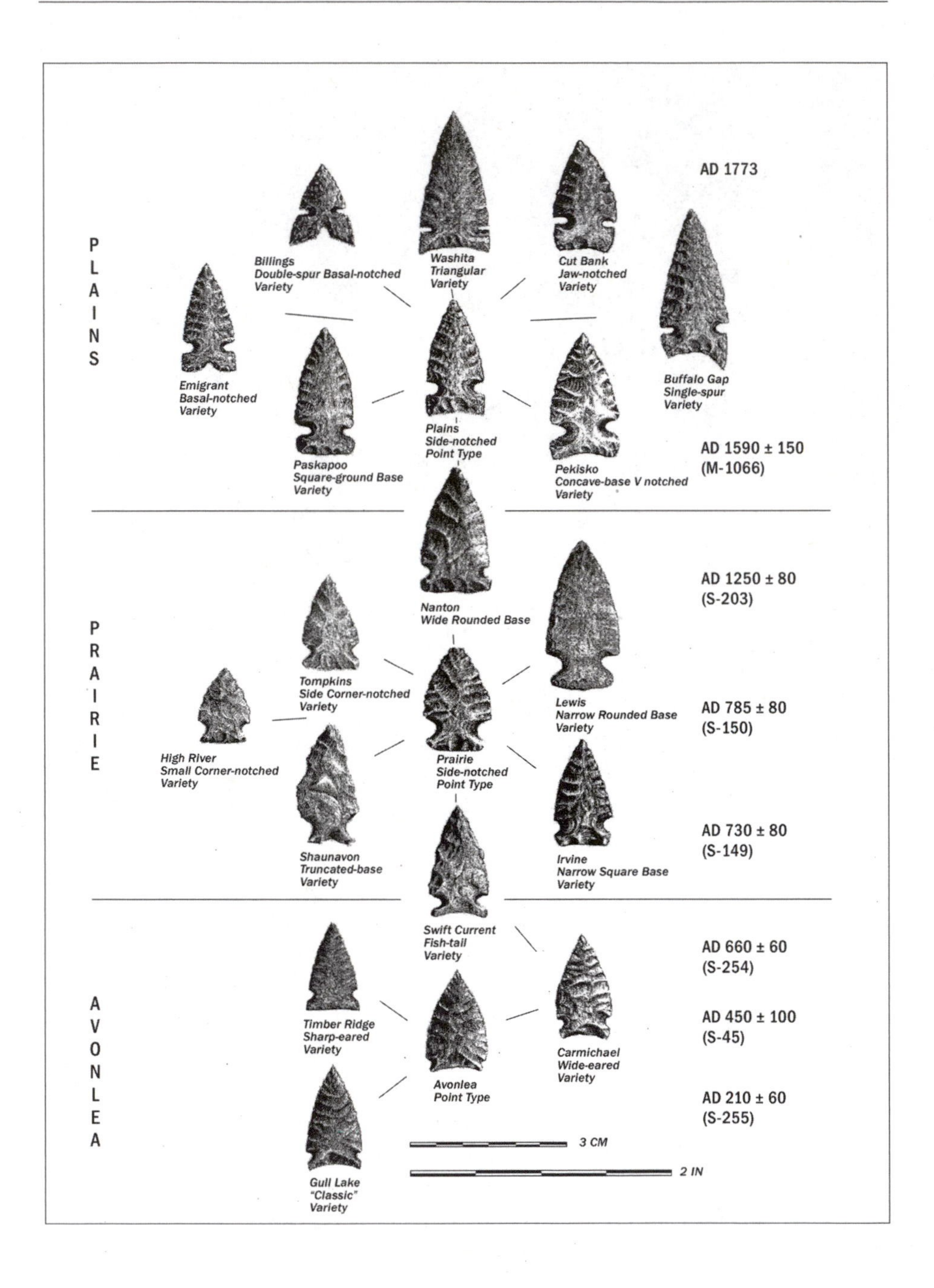

FIGURE 2.4 Sequence of styles of stone points, Gull Lake site. *Drawing by Ronald Sager and Gail A. Ebert for Milwaukee Public Museum, The Gull Lake Site.*

Avonlea style *and* the use of arrows and (inferred) bows were new in the northern plains at this time. Evidence *against* this hypothesis is negative: the lack of dense butchered bone below the Avonlea layers. However, in archaeology, failing to find something does not prove that what you were looking for didn't exist, or even that it didn't exist in the site you're investigating: the *probability* that an archaeologist today digs exactly where a particular artifact lies is not high, and it is always possible that the diagnostic artifacts were carried away from the location by the people who used them. We can never see the entire real past.

To outline the scientific method as used at Gull Lake:

- Begin with collecting observations, that is, data. For archaeological excavation, first map the site and lay out a grid so that every observation (every datum) can be recorded precisely for its three-dimensional location in context with associated data. Then uncover what is buried, proceeding as slowly as time allows for the project, leaving every datum in place until recorded photographically, in words, and on a map. Today, archaeologists also use remote-sensing equipment such as magnetometers and ground-penetrating radar to discover buried phenomena, and geographic information systems (GIS) for mapping data.

- Step two, classify collected data. Records, artifacts, and associated natural objects and data such as butchered bone, seeds, and soil samples are brought into the laboratory to be examined and to be compared with artifacts within the collection and from other sites. Experiments may be made with replica artifacts to better determine how excavated ones were used. Specialists in other fields, ranging from zoologists and soil scientists to local farmers and historians who know a region intimately, are often consulted. When everything has been meticulously examined and identified, like is placed with like and differences between groupings are set out.

- Step three, formulate an interpretation of the site, or of selected data from the site. We had a great deal of data supporting our working hypothesis that Gull Lake held a series of pounds into which bison herds had been driven, to be slaughtered as the staple food of Plains Indian communities. Gull Lake didn't preserve corral poles, so far as our excavations went, but the layout of a semicircle of dense accumulation of bison bone, with little outside the semicircle, corresponds to the semicircle with fallen poles at the Blackfeet site. Inference to the best explanation, at the time, was that people using bows and arrows and driving game animals into pounds for slaughter had migrated southeastward into the Canadian prairies around two thousand years ago—southeastward from the forests of northwest Canada because neither Avonlea-style points nor bison pounds earlier than two thousand years ago were then known east of Saskatchewan.

- Step four, write up your work and make it available to other researchers. Get your colleagues' feedback and questions. Reexamine your hypothesis based on the feedback. Science depends upon researchers sharing data and discussing interpretations.

When he wrote *The Gull Lake Site*, Tom Kehoe had my assistance in striving for precision and clarity, and then gave the manuscript to a "wordsmith" (that's what she called herself) to edit. Our wordsmith told us that every sentence ought to be phrased in more than one way, and then the clearest and easiest-to-read version selected. It was a time-consuming process, but it resulted in a monograph still cited and discussed a generation later. The capstone of the research was a major research paper by an archaeologist of that younger generation, bringing substantial newer data to bear on the Gull Lake interpretation. Now, we learned, Avonlea-style artifacts have been found east of Saskatchewan in Manitoba and Minnesota, in sites at least as old as the Avonlea occupations at Gull Lake, while no Avonlea style has yet been found in northwest Canada. Our interpretation has been refined to acknowledge that Avonlea-style arrowpoints may represent the introduction of bow-and-arrow weapons into the northern plains, but the varieties within that style, and of sites they appear in, suggest a technological innovation spreading across several ethnic groups. Our younger colleague, in true scientific spirit, acknowledged our interpretation had been reasonable until widespread research produced many more data, making inference from the range of newer data modify our explanation derived from fewer sites.

■ ■ ■ EXPANDING RESEARCH: THE WIDER QUESTION ABOUT BISON DRIVES

Archaeologists working in the northwestern plains, and Blackfoot and other First Nations whose territories lie in the region, wanted to know how old the practice was of driving herds into pounds. The same question arises in Europe regarding Paleolithic hunters: Did they drive herds into corrals, and if so, when was the practice invented? Many Upper Paleolithic sites (during the last part of the Pleistocene ["Ice Age"] epoch, from around 35,000 years ago to about 9,000 years ago) contain butchered bone from herd game such as bison, horses, wild cattle, and elk. Some sites lie under bluffs, with butchered bones clustered the way they are in American bison pounds. Representations of these same animals are well known from cave walls in France and Spain, with enigmatic lines, boxes, and dots around or beside the prey. At Lascaux Cave, considered perhaps the greatest of the Paleolithic art sites in France, one wall has a line of horses going to an edge and an upside-down horse over the edge, as if falling. Tom Kehoe expanded his research on game animal drives by examining European sites to see if his hypothesis about their use applied in Europe as well.

Three substantial bison pound excavation projects, and numerous visits to other bison drive sites excavated by colleagues or brought to his notice by avocational archaeologists, gave Kehoe experience to evaluate European data. His first venture was for us to work as volunteers for six weeks at the site of Solutré in eastern France (Figure 2.5), where quantities of butchered wild horse and reindeer

FIGURE 2.5 Solutré, France, excavation of site with horse and reindeer drives, ca. 15,000 B.C.E. *Photo by Alice B. Kehoe.*

bones lay at the base of a limestone outcrop. We observed that the horse slaughter remains clustered like the bison bones we were so familiar with, and lay directly under a cleft in the outcrop that would have funneled down a herd grazing on the grassy upper slope, if it were startled and driven toward the hidden break. It seems probable that around sixteen or seventeen thousand years ago, long before our American hunters, Paleolithic hunters had practiced the technique of driving herd game over bluffs into corrals.

More controversial is Tom Kehoe's suggestion that rectangles and lines of dots close to herd game depicted in cave art represent traps, specifically drive lanes and pounds or deer fences (long nets strung between trees, into which deer historically were driven). Someone familiar with maps of bison drives easily sees the figures and series of dots as similar depictions. European archaeologists lacking such experience are not easily persuaded. American Indians occasionally pictured game animals and even lines of herd game as if driven, but nothing like the amazingly realistic animal murals in the European caves is known from the Americas; Indian rock art of the most recent centuries favors battle scenes, warriors, and spirit beings guarding nations' territories and routes. Tom Kehoe's interpretation of the cave art he cites is inference to the best explanation, given his experience in North America and Europe, while European archaeologists' skepticism reflects their different experience—they, too, practice IBE. Using scientific method doesn't automatically produce a consensus on what is the best explanation.

Practically all archaeologists will tell you they follow scientific method. Empirical observation, precise recording, measuring, classifying by observable criteria, drawing upon ethnographic and historical descriptions of human behavior and its material consequences to interpret what is represented in a site, make archaeology a science. Certain models taken from history or the natural sciences, such as animal population adaptations to environments, will be favored "paradigms," to use Kuhn's term, for interpretation, to be superseded after some years as a younger generation looks critically at its predecessors. Although this book is about controversies in archaeology, there is, overall, consensus on the necessity of carefully using scientific method and understanding what scientific method can, and cannot, do. Archaeologists can call upon the humanities for additional, or broader, interpretive models, making clear why one or another model or case example seems appropriate for the archaeological data. There is a larger world out beyond archaeologists' conferences, a world of people wanting exciting mysteries and drama. We'll look into "popular archaeology" in the next chapter.

SOURCES

Barnes, Barry, David Bloor, and John Henry, 1996. *Scientific Knowledge: A Sociological Analysis*. Chicago: University of Chicago Press.

Lakoff, George, and Mark Johnson, 1980. *Metaphors We Live By*. Chicago: University of Chicago Press.

Kehoe, Alice Beck, 2004. When Theoretical Models Trump Empirical Validity, Real People Suffer. *Anthropology News* 45 (4):10.

Kehoe, Thomas F., 1973. *The Gull Lake Site: A Prehistoric Bison Drive Site in Southwestern Saskatchewan*. Publications in Anthropology and History No. 1. Milwaukee, WI: Milwaukee Public Museum.

—. 1990. Corralling Life. In *The Life of Symbols*, edited by Mary LeCron Foster and Lucy Jayne Boscharow, pp. 175–193. Boulder, CO: Westview Press.

Kelley, Jane H., and Marsha P. Hanen, 1988. *Archaeology and the Methodology of Science*. Albuquerque: University of New Mexico Press.

Kuhn, Thomas S., 1962. *The Structure of Scientific Revolutions* (2nd edition 1970).Chicago: University of Chicago Press.

Quinn, David B., 1974. *England and the Discovery of America, 1481–1620*, pp. 22–23. London: Allen and Unwin.

Schmandt-Besserat, Denise, 1992. *Before Writing: From Counting to Cuneiform*. Austin: University of Texas Press. (For more on the development of writing, with discussion of controversies, see Stephen D. Houston, editor, 2004, *The First Writing: Script Invention as History and Process*. Cambridge: Cambridge University Press.)

Shapin, Steven, and Simon Schaffer, 1985. *Leviathan and the Air-Pump: Hobbes, Boyle and the Experimental Life*. Princeton: Princeton University Press.

Walde, Dale A., 2006. Avonlea and Athabaskan Migrations: A Reconsideration. *Plains Anthropologist* 51 (198):185–197.

3 Popular Archaeology

Archaeology is exciting. The Temple of Doom! The Curse of the Pharaoh! Tombs! Gold! Indiana Jones, machete in hand, breaking out of the jungle into a fabled city. Lissome, pistol-toting Lara Croft. Archaeologists are sexy, brilliant, quirky characters like on Crime Scene Investigation.

Even archaeologists can daydream.

In reality, your archaeology instructor probably doesn't look or act much different from your economics or chemistry professor. Lives in a modest house with spouse, kids, dog, cat. Real field archaeology is plodding work, day after day scraping dirt with a trowel, stopping often to measure precisely and record the figures. On wet days, archaeologists come indoors to wash the little broken pieces of cooking pots and stone blades, writing catalog numbers on them and sorting. An archaeologist can be literally a bean-counter, studying an ancient community's food remains. We archaeologists can get excited as data build into patterns, reveal contacts and social change, but it's intellectual excitement. Real archaeologists do their damnedest to avoid any kind of danger.

Not everybody accepts archaeologists' polite explanations that real work is seldom glamorous. It's like talking about UFOs: "There's a cover-up. You're not telling us." Rumor starts in the local community. "They found this treasure and took it out at night, when the workmen had gone home." After archaeologists leave at the end of a season, people may come in with pick-axes, shovels, maybe backhoes, tearing up the site to find more treasure; nobody's seen any treasure in the archaeologists' camp, but of course, they're sure it was illegally sneaked away. High prices collectors pay for well-chipped stone points and whole pots fuel suspicion and stimulate looting at sites.

■ ■ ■ SPACEMEN AND OTHER FRAUDS

Plenty of people are willing to accept supernatural or extraterrestrial angles to archaeology. Hobbyist investigators of the most famous UFO sighting, in 1947 near Roswell, New Mexico,[1] heard that at the time, an archaeologist had been in the field north of Roswell. Meticulously they combed archaeological permits on file in the state, turning up the project and director's name. The amateur investigators tracked it to a retired scientist in West Texas, and telephoned him. He was 95 years old, and as his wife told them, his memory was failing. Should his statement that he saw nothing there in 1947 be taken as a negative, or the result of memory loss? If only they had managed to find him earlier! In fact, the elderly man had taken out the permit for a project carried out by a graduate student; that person had been in the field, not the professor. The student, now advanced in a career in archaeology, enjoys assuring listeners that no one working on her Capitán Valley site saw or heard anything unusual during the summer of 1947.

The most famous of the extraterrestrial apologists, Erich von Däniken, a Swiss, wrote a 1970 bestseller, *Chariots of the Gods*, identifying various famous archaeological sites as constructions designed by extraterrestrial travelers to Earth. *Chariots of the Gods* became a phenomenon in its time, followed by almost thirty other books, a box-office smash documentary film, and even a theme park in Switzerland. Von Däniken has sold over sixty million books, translated into thirty-two languages, with his claim that ancient history was driven by spacemen from another planet. Among his cases is the bas-relief lid of the stone sarcophagus in the tomb of Mayan king Pacal of Palenque (A.D. 603–683), in southern Mexico.[2] The beautiful carving shows the king rising toward heaven, escaping the jaws of the underworld monster (Figure 3.1). The Bird of Heaven awaits him, sitting on the World Tree. Around the sides of the sarcophagus, a text in Mayan hieroglyphs gives Pacal's royal genealogy, his parents and their noble origins. Despite scholars' translations of the text and well-known Maya iconography, von Däniken claimed the picture shows an extraterrestrial in his spaceship. How does von Däniken know that? He has psychic powers, he says. Thanks to these powers, his spirit travels the world and universe

---

[1] This is where what appears to some to be bodies of extraterrestrials were seen on the ground. Most of us notice that the bodies look like crash-test dummies, and classified military experiments with proto–space capsules were carried on in the Southwest during the period.

[2] See the award-winning *Ancient Mexico and Central America* by Susan Toby Evans, pp. 329–335, for vivid text and pictures of Pacal. The real archaeology Evans shows is truly exciting.

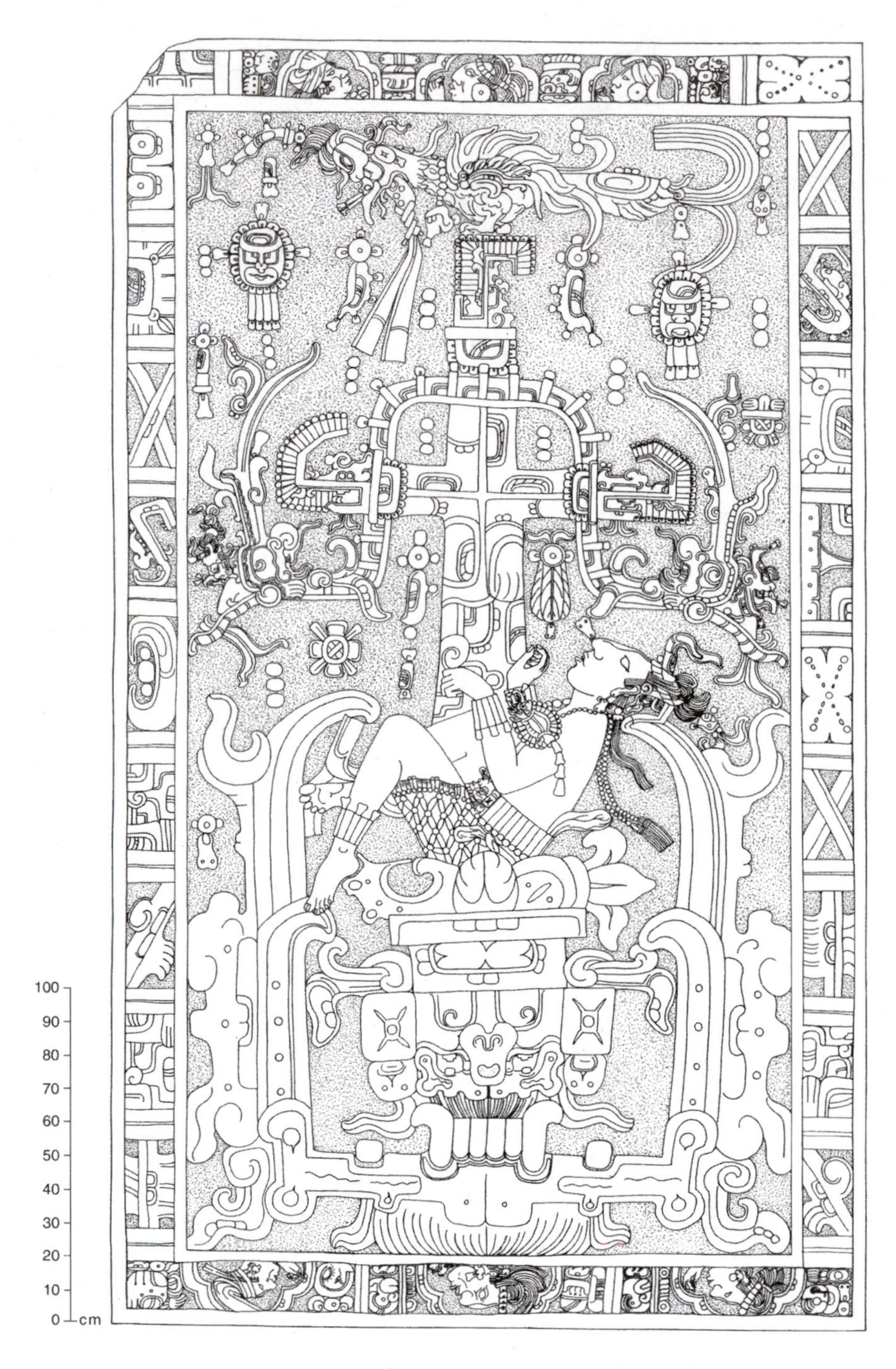

FIGURE 3.1 The cover of the sarcophagus in tomb of Lord Pacal of Palenque. Pseudo-archaeologist Erich von Däniken was convinced it was the representation of an alien spaceship. © *Merle Greene Robertson, used with permission.*

discovering how stupid scholars are. In another instance, he told the *National Enquirer* tabloid that a construction of small boulders on a Canadian prairie hilltop, published in the *National Geographic*, is a landing pad for spaceships. The *Enquirer* had a staff member telephone the archaeologists who had mapped and excavated the site—Tom and Alice Kehoe (see Chapter 8)—and when we insisted that the construction was prehistoric Indians' means of keeping a calendar by noting the summer solstice, the reporter argued for von Däniken's explanation. "But look, we spent two seasons working on the site. We interviewed Indian people from every one of the prairie First Nations about the site and how it might have been built and used," we protested. "Von Däniken has never been there." "Yes, he had," the reporter countered. "He sent his spirit and it was there for fifteen minutes. It saw everything."

Von Däniken needed to send his spirit to Canada because his body was in a Swiss prison, convicted of fraud. Not about the book, just ordinary shady dealings with money. *Chariots of the Gods* capitalized on the popularity of "UFO mysteries," fed by occasional odd electrical discharges, reflected images, space junk (bits of meteorites, pieces of broken unmanned satellites and spacecraft), and stories by people convinced they were abducted by aliens while they slept. When American astronaut Neil Armstrong really did walk on the moon, in the summer of 1969, it seemed more plausible that the reverse had happened, that is, beings coming from outer space to walk on Earth. Taxpayers saw millions of their dollars spent to build space capsules; outer space is our frontier for exploration. Attributing strange-looking sites and art from other cultural traditions to extraterrestrials fits into supporting NASA projects to explore the extraterrestrial universe. The problem for archaeologists is that von Däniken and his imitators expropriate actual sites and antiquities, denying credit to the peoples such as the Maya who did create them. In a sense, von Däniken stole the achievements of Pacal's Maya citizens and of the ancestors of the Saskatchewan Indian people who scientifically mapped out in stone the positions of six astronomical bodies at solstice.

Though science cannot disprove the existence of spacemen, nor that they visited earth, our methods of archaeological science have not produced any evidence that would lead a rational archaeologist to believe this interpretation of ancient history. Ockham's Razor—the simplest explanation is the best—tells us we should look for terrestrial explanations for ancient phenomena before we begin to invoke spacemen.

There is also an American purveyor of outrageous archaeology who had been in a penitentiary. When he was released, he teamed with a former prison guard to promote golden treasures and mysterious stone tablets supposedly from a newly discovered cave in the hills of southern Illinois near the penitentiary. The two held public auctions of the stuff, and amazingly, people paid good money for "gold ornaments" without getting them appraised. A geologist checked the tablets and found they were made of a stone sold commercially for lithographs. Archaeologists and other citizens ask to see the cavern, but promises to take them have never materialized. In spite of years of this flim-flan, a few still want to believe that a dynasty of Asian kings lies in splendid tombs in a cave in southern Illinois.

With such a slew of sleight-of-mind con artists and quacks around, a few professional archaeologists have succumbed to temptation. Most notorious was Professor Frank Hibben of the University of New Mexico (1910–2002). He impressed everyone with his rugged good looks, deep voice, and brilliant mind. In 1936, as a graduate student at the University of New Mexico, he married a wealthy woman and, with several fellow students, began excavating a cave in the Sandia Mountains outside Albuquerque. Five years later, he made *Time* magazine with the announcement that he had discovered the oldest artifacts in America, twice as old as Folsom (that is to say, twenty thousand years old). Distinctive spearpoints resembling European Upper Paleolithic Solutrean spearpoints lay with mammoth, mastodon, and bison under a layer colored yellow by mineral ocher, above which were Folsom points sealed under desert hardpan crust. Next, he chartered a boat to explore Alaska and found Folsom points and mammoth remains on a remote beach. After serving in World War II, Hibben returned to Albuquerque to become the most popular professor at the University of New Mexico, good friend of politicians who enjoyed his hospitality and shooting excursions, and, according to his own stories, secret agent in the mountains of China.When radiocarbon dating was introduced at the end of the 1940s, Hibben sent samples from Sandia Cave and published dates apparently confirming his prewar estimate of a Solutrean age, seventeen thousand years old, for the Sandia points layer. A rancher brought him out to see a site that produced more fine Sandia points and mammoth bones. Only, no one else ever found Sandia points! Dozens of archaeologists working in Paleoindian sites across North America found Folsom, Clovis, butchered mammoths, controversial "pre-Clovis" habitations, but never a Sandia-style point. Several geologists and archaeologists revisited Sandia Cave and could not see either a yellow ocher layer or a hardpan; instead, they found the cave riddled with packrat burrows. Other inconsistencies were noticed with what Hibben said he had sent for radiocarbon dating and what lab workers recorded, and with the remote beach in Alaska, revisited by a team of archaeologists and geologists who found no evidence of any artifacts or mammoths. Curiously, Hibben's Sandia Cave excavation notes and the collections in the university museum lacked the debitage—flakes chipped off stone artifacts, broken bits—usual in occupation sites. Did Frank Hibben fake Sandia Cave? What about his adventures on safari? and as a secret agent? However this charming and generous man wiggled around unpleasant questions, it is clear now that he steadily misrepresented Sandia Cave.

■ ■ ■ PYRAMIDS AND MOUNDS

Pyramid magic is a perennial fantasy that fascinates the public. Why Egyptian royalty in the third millennium B.C.E. chose to order enormous pyramids of stone blocks constructed to enclose their tombs, we don't really know. Their priests and scholars were developing sophisticated mathematics (Classical Greeks went to university in Egypt), and the pure geometry of the great pyramids at Giza proclaims not only physical but also intellectual power.[3] During the eighteenth-century European Enlightenment era, heyday of rational logic, counter-movement theories sprang up that pyramid measurements encoded hidden messages, or foretold the future, or the shape itself had mystical power. Mozart's opera from this time, *The Magic Flute*, has its hero, Prince Tamino, initiated into a brotherhood of wise men through a ritual inside a pyramid. In the 1970s, "pyramid power" was a fad, people made little plastic-sheet pyramids and put food under them to make the food potent. (My kids tried this. I can tell you that slices of pizza shrivel up under plastic pyramids.) Pyramids do have potency to sell books. Pyramids also are used to buy books, literally: on the back of the U.S. dollar bill is a magical pyramid topped by an all-seeing eye. George Washington, on the other side of the bill, like Mozart, was a Freemason, whose Egyptian-tinged symbols continue in today's Masonic lodges.

William Flinders Petrie (Figure 3.2) was a British youth in the 1860s who read about alleged predictions of the future that could be calculated from Giza pyramid measurements. He learned how to survey architecture accurately, went to Egypt as a young man, and soon was disillusioned by finding that the book that had impressed him had not used careful measurement (see Chapter 2 on precise measurement as necessary for scientific method).[4] Flinders Petrie devoted his

---

[3] See Hornung's book *The Secret Lore of Egypt* for supposedly Egyptian fantasies and symbols. Hornung has also written a good introductory *History of Ancient Egypt*.

[4] See Paul Bahn's *Cambridge Illustrated History of Archaeology*, pp. 148–149, 160, for Flinders Petrie's path-breaking work.

FIGURE 3.2
(ABOVE): W. Flinders Petrie, leading Egyptologist of the late nineteenth century who introduced seriation in analyzing artifacts. *Courtesy Petrie Museum, University College London. (RIGHT):* Tom Kehoe, contemporary archaeologist, pointing with his trowel to a stratum in a bison drive site. The layers of remains separated by soil layers are the stratigraphy of the site, the basis for seriation of artifacts. *Photo Alice B. Kehoe.*

life to painstaking excavations, establishing sound chronologies of sites and artifact styles. He initiated the method of "seriation," noting how pottery styles changed over time, linking styles to dated tombs, and then using styles to date sites without inscriptions. Hundreds of archaeologists from many nations, following Petrie's scientific methods, have provided detailed knowledge of Egyptian history covering thousands of years (see Chapter 8 for more on pyramids). None of this mass of information seems to deter lovers of mystery from speculating about the "secrets of the pharaohs," including the tantalizing idea that extraterrestrials started it all.

Americans have constructed prehistoric pyramids not at all like those at Giza. Throughout the agricultural zones of the Americas, from Peru through the United States, flat-topped (truncated) pyramids were built to raise temples and palaces above the common throng. Rounded mounds were built over tombs. None of the American structures were piles of stone blocks like the Egyptian pyramids; instead, American First Nations built with adobe bricks (millions in some Peruvian platform mounds), or cores of rubble stone veneered with finely cut slabs, or of earth engineered with drainage layers inside to prevent collapse. The oldest known American mounds (in Louisiana), earth-built, are as old as Egyptian pyramids. U.S. antiquarians in the nineteenth century argued whether American Indians could have built these impressive mounds, upon which question hinged evaluation of the "civilization" of First Nations. Advocating conquest and forced assimilation to Euroamerican culture, popular sentiment attributed sites to a "race" of Moundbuilders destroyed by savage invading Indians (rather like invading Euroamericans destroying indigenous Indian settlements!). The Smithsonian's Bureau of American Ethnology decided to examine the question in the 1880s by a broad-scale research project. Its conclusion was that American mounds could have been built by ancestors of contemporary American Indians; no evidence was found for the hypothesis of an earlier "race" extinct by the time of Columbus. This conclusion did not refute popular prejudice, but merely led opinion to dismiss mounds as nothing more than piles of earth—savages' structures.

To build a rectangular earth mound one hundred feet high with a flat top larger than a football field that retains its shape for more than a thousand years, such as the spectacular mound at Cahokia across the river from St. Louis, Missouri, is in fact an impressive engineering feat (Figure 3.3).[5] Cahokia itself,

[5] See "Sources" at the end of this chapter to find information on books by Timothy Pauketat, who has written extensively on the greatest ancient city in the United States.

FIGURE 3.3 Cahokia's principal mound, called Monks Mound, is one thousand feet long and one hundred feet high.

with over a hundred mounds and a Grand Plaza built level one-third of a mile long, was obviously the capital of a state, although one that held power for only a couple of centuries before breaking up into small kingdoms. The only real city in prehistoric United States, Cahokia has been designated a UNESCO World Heritage Site, but the state of Illinois, which owns the huge site, was slow to protect it, and few Americans have heard of it. The myth of America the wilderness inhabited by savages has blinded people to the extent that they can't see an architectural wonder a hundred feet high, one thousand feet long, and seven hundred feet wide—Cahokia's principal platform mound. In Wisconsin, there's a Pyramid Motel built to look like an Egyptian pyramid, and not many miles away, there's a thousand-year-old town site called Aztalan with two American-style flat-topped pyramids. American pyramidal mounds haven't been tagged "mysterious," although we know less about the thousands of great pyramidal mounds here than we do about the relatively few pyramids in Egypt.

Like the ones in Egypt or the American Midwest, pyramids are common features of many ancient societies. China, Iraq, Peru, Mexico, and many other places have pyramids, an effective design for building a large, impressive structure that will survive erosion, earthquakes, and invaders. They resemble the mountains held sacred by many cultures, impressing citizens with the power of rulers and priests who built and had access to them. Some pyramids are built of stone, some of brick, some of rubble, some of earth. Some are used as tombs, some as foundations for temples or palaces. Some were built five thousand years ago, some thousands of years later. Talking about pyramids as if they were a single, unified phenomenon ignores the diversity of contexts for these structures. What archaeologists can say is that they are among the most impressive engineering achievements of the ancient world, and that they served a variety of functions. But we won't claim them to have supernatural powers or special preservation capabilities for yesterday's pizza. We stick with the evidence.

■ ■ ■ PSYCHIC ARCHAEOLOGY

Digging is destructive, digging is time-consuming. Why dig a site if there's a faster, nondestructive, and more informative way of learning about the past at a locality? Employ a psychic! That perennial television gimmick hangs on the edge of archaeology. A person claiming extrasensory perception (ESP) may be asked to sit at a known site and tell about its inhabitants' behavior long ago, or taken to an area to recommend where to excavate, or given an object to identify. Almost always, the "experiment" is judged successful, with fulfillment attributed to *psi* (the Greek letter ψ) capability, a tenuous ability to perceive more than most other people can. "Psi" was coined by researchers J. B. and Louisa Rhine in the 1920s to assert their investigations premised some physically real quality, not spirits.

Unquestionably, some people enjoy heightened perceptive ability compared with ordinary humans. Musically gifted people hear pitch, notes, overtones, that the untalented cannot distinguish—my mother and I just gaped as my musician father and brother listened and noticed so much that we simply could not hear. Experience or training can increase sensitivity, as in professional tasters. Farmers can tell a lot about soils, mechanics and doctors interpret subtle noises, and animals are informed by sounds and smells we can't pick up. It also makes a difference, as we all appreciate, to concentrate, focus intently, on a particular experience.

Psychic archaeology seems to have capitalized on several different abilities: local residents' deep familiarity with a place; certain persons' unusually acute perception and recognition; and other individuals' willingness to spin a good yarn. That latter performance isn't necessarily deliberate; Frank Hibben's friends and students judged that he got carried away by his stories, and elderly people may lapse into filling in or enhancing memories. My mother (again) in her nineties told wonderful tales of her childhood, of watching the first electric lights and Lindbergh crossing the Atlantic, oblivious to her grandchildren doing a bit of mental arithmetic and concluding that she couldn't have been ten years old for each event!

Accounts of psychic archaeology tend to claim that the academic archaeologist is amazed by the unlettered psychic's uncanny ability to tell the professor where to dig to find things, or to identify a fragmentary artifact or feature. Most of what I've read doesn't credit the uneducated person with their lifetime of direct experience working in the landscape and with their hands. It wasn't psychic ability that prompted one of my excavation crew, at a fur trade post site, to tell me that the fine soil we were troweling was produced by someone sweeping the dirt floor with a homemade straw broom; Reinhard's mother had done that in their homesteading cabin, and he remembered being a little boy sitting on such a dirt floor. Older archaeologists who had grown up on farms have been legendary for analyzing soil by tasting it. Some archaeologists can tell the difference between different pottery styles by taste or sound, rather than sight. An account of camping with Blackfoot Indians in the 1890s marveled at how the elderly matron of the band slowly selected spots for their tipis; after a heavy rain overnight, these spots remained dry because her practiced eye had noted small differences in elevation on the grassy prairie. If you've seen enough Plains bison drive sites, as I have, you too can point to the section of the valley rimrock where there's likely to have been a drive and pound; you recognize the configuration needed to get a herd down a bluff into a corral. When you write up your survey report, you articulate what had begun to feel intuitive.

As for the storytellers who in half an hour will vividly describe life in the ancient pueblo, sparing us the cost and discomfort of digging: a willing suspension of disbelief is the ticket to enjoying fiction, but not to science. Some of their picture will be common sense—"Mothers are nursing babies"; some will be observation of traces of ruined walls, plazas, steps; some will be reasonable speculation—"Hunters are carrying in a deer"; and some seem stereotype—"Mothers are teaching daughters to make pots." None of the story requires *psi*, whatever that may be; portions might be verified by excavation, and the rest cannot be checked. Their pronouncements fall outside the range of science's ability to confirm or debunk. Psychics may sincerely believe in their extrasensory power, but faith doesn't make it so.

No matter how attractive they sound to the general public, psychic time travel, extraterrestrials, pyramid power, secret codes, and lost cities defy reasoned efforts to explain how much we do know about the human past. They persist, like the urban legend that someone threw a baby alligator into a city sewer, it lived and grew, and one day swam up into a house plumbing pipe and bit an innocent citizen sitting upon the toilet.

■ ■ ■ ATLANTIS

Back in the fourth century B.C.E, the Greek philosopher Plato wrote a parable about a great island destroyed by cataclysms. Parables—little stories teaching a moral—were a favored literary form then, and for several centuries as well, as you can see in the New Testament's parables attributed to Jesus. Later, Plato's story, meant to teach that natural forces can overwhelm human endeavors, was mistaken for history. Readers unfamiliar with literary conventions took for a fact the existence of a fabulous civilization called Atlantis, in the ocean just past the Straits of Gibraltar, the outlet of the Mediterranean. Supposedly, Athens's revered statesman Solon had gone to Egypt to study, two centuries earlier than Plato, and there learned priests had told him that Athens—Solon's city—had in ancient times, nine thousand years earlier, been destroyed "like Atlantis." Greeks believed that Egypt was the oldest nation, with a store of knowledge far surpassing that in Greek cities; Solon had been ignorant of Athens's long-ago glories and how, he was told, her army had held the eastern Mediterranean lands against Atlantis's invasion. The point of the story was that earthquakes and floods swallowed up both brave Athens and mighty Atlantis, leaving in Greece only simple shepherds, and where island Atlantis had been, the ocean's waves. Egypt alone preserved civilization, and Plato's Athens considered itself fortunate that Solon had founded its institutions with Egyptian counsel.

Archaeologists have debated whether Plato repeated a thousand-year-old story of the explosion of a volcanic island, Santorini (also known as Thera), about 1628 B.C.E. This island was in the eastern Mediterranean, near Greece. Half the island sank beneath the waves because of the eruption, and its blanketing ash and tsunami floods wrecked Bronze Age cities along the region's coasts. Considering the gap in time between this eruption and Plato, and that many volcanic eruptions and floods occurred during those thousand-plus years, there's not much point in guessing whether Plato meant Thera. More to the point is that Thera was not beyond Gibraltar in the Atlantic; it never ruled most of the Mediterranean, nor raised a huge army attacking Athens. If there is any historical element in

Plato's parable, it might have come from our knowledge of Phoenician merchant ships sailing through the Straits of Gibraltar into the Atlantic. No firm evidence of Phoenician voyages across the Atlantic to the mid-ocean Canaries and Azores islands, or to America, has been recognized, although a little information may have survived among medieval Arab scholars. Trying to get archaeological evidence for Atlantis misses Plato's purpose in writing the parable: to exemplify the moral that natural disasters can destroy both mighty invaders and noble small states.

Though never intended as a historical story, Atlantis became a legend that can't be knocked down. Artur Posnansky, an Austrian emigrant to Bolivia early in the twentieth century, used his civil engineering skills to map the pre-Inca city of Tiwanaku in the Andean highlands near Lake Titicaca. He noticed that several alignments of monumental structures could have been used for astronomical observations, either around 11,000 B.C.E, or around A.D. 1000. If Posnansky had published the latter date, he would be hailed as a pioneer archaeologist, for Tiwanaku does date between A.D. 500 and 1150. Unhappily, he was thrilled by the alternate date of 11,000 B.C.E.: Tiwanaku must be Plato's Atlantis, it's thousands of years older than Plato! It's beyond the Straits of Gibraltar! (Thousands of miles beyond.) No matter that the city lies twelve thousand feet high in the middle of a continent, that it never was an island (unless you count the moat built around its center), and it wasn't destroyed by either earthquake or flood. Posnansky's elaborate history had to ignore Plato's lines about his Athenian ancestors repulsing the Atlantis army. Missing what really mattered to Plato is typical of the Atlantis-myth buffs.

Posnansky's Atlantis is only one of over a hundred locations on every continent (including Antarctica!) that has been proposed as Atlantis over the past centuries (Figure 3.4). Each claim takes various details from Plato and tries to fit a specific geographical location to his story. None will ever match all the key elements, for Plato described a mighty, technologically sophisticated city-state dating to 10,000 B.C.E., when in fact only the first experiments in animal domestication, crop planting, and moving into small villages were being tried out in a few places on earth, and most human populations were living as hunters and gatherers in small family bands. The best inference, then, is that Atlantis was the fanciful subject of a parable, not a lost continent begging to be found.

■ ■ ■ ATLANTIS AS A POP CULTURE MAGNET

In 1996, your author participated in a television film titled "In Search of Atlantis." The producer explained that if she had "Atlantis" in the title, fourteen

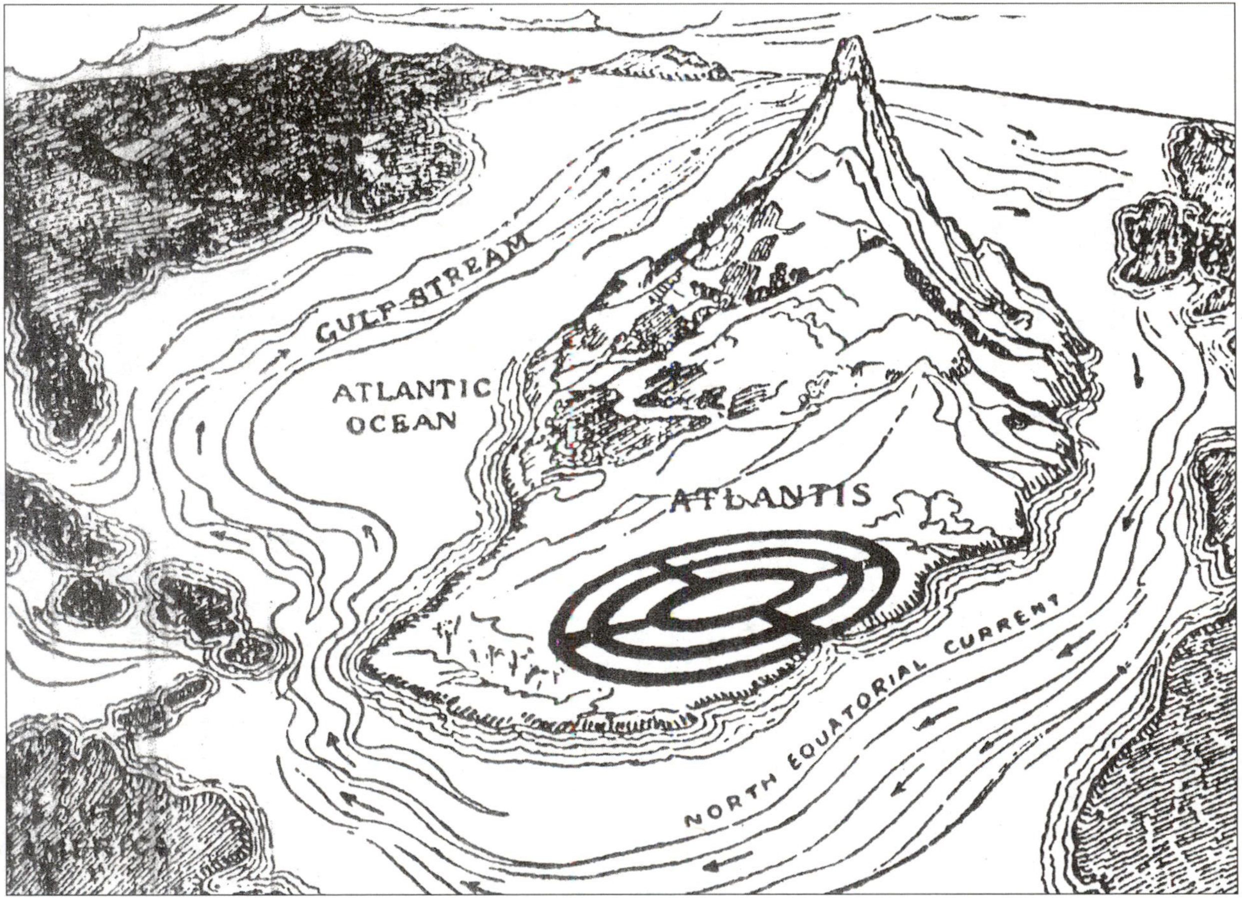

FIGURE 3.4 Speculative drawing of Atlantis, by Paul Schliemnn (1912).

million people would watch it. With a more accurate title such as "Visits to Fascinating Archaeological Sites—Hissarlik, Olmec San Lorenzo in Mexico, and Tiwanaku," the film would bomb. "Atlantis" was the hook, briefly introduced at the beginning and then used as the question, "Could this site be Atlantis?" Hissarlik is the Turkish site generally accepted as Homer's Troy, where the Trojan War took place, and the nice British archaeologist working there calmly explained how it has nothing in common with Plato's Atlantis—it's on the opposite end of the Mediterranean from the Straits of Gibraltar. Tiwanaku and San Lorenzo Olmec sites are west of the Straits, very far west, being in America west of the Atlantic Ocean, and the Olmec, 1500–400 B.C.E., date earlier than Plato, but like Tiwanaku, their towns were not destroyed by earthquakes and floods; they merely fell into ruin after political power shifted elsewhere. Fourteen million people watching the Discovery Channel learned that none of these sites could have been Atlantis, if Atlantis had existed.

For several years as the Discovery Channel reran "In Search of Atlantis," I'd get telephone calls and letters telling me about more Atlantis candidates. My favorite is a scenario detailing how royals had fled across the Atlantic to Mexico, ordered the Indians to build cities with pyramids—now called "Olmec"—and when after some centuries the Olmec people rebelled, the dynasty fled again, across the Gulf of Mexico. They went up the Mississippi River to the Arkansas River, said this correspondent, at last building a new city near Hot Springs, Arkansas. The gentleman enclosed a U.S.G.S. topographic map including the Hot Springs Country Club, to show me buried walls of Atlantis visible as ridges on the golf course. Dedicated scientist though I am, I haven't taken the time to drive to Arkansas and test those embankments. I'll stick with an alternate working hypothesis that the topographic map shows features built for the golf course.

Another letter invited me to join an expedition to explore an alleged canal linking Tiwanaku to Lake Poopo and a route eastward to the Amazon River, the Atlantic, and across the ocean to India. This correspondent accepted that Tiwanaku was Atlantis (so had to be linked somehow into Plato's Mediterranean-India trading world). For several reasons, I did not accept the invitation to the expedition. First, I would have to pay my own way, five thousand dollars at least, and the schedule would require me to take a semester's leave, without pay, from teaching. More crucial was the fact that I had worked at Tiwanaku and knew that the supposed canal is the Desaguadero River. Desaguadero is Spanish for "without water." The river is clearly a natural channel through a mountain pass giving outlet to Lake Titicaca, ending in Lake Poopo. For much of the year, practically no water flows through this channel. My correspondent, Colonel

Blashford-Snell of Dorset, England, did take an expedition down the Desaguadero in 1998, towing reed boats with jeeps because there wasn't enough water to float even reed boats. The Colonel then realized that Poopo has no outlet over the mountain rim into the Amazon lowlands, so the next year he brought his reed boats down by railroad. Uncooperative geography notwithstanding, Blashford-Snell sailed his reed boats and modern back-up during several more seasons along lowland rivers, intending in 2005 to sail around southern Africa's Cape of Good Hope to the Persian Gulf and Red Sea. That intention was the last posting on his web site when I accessed it in January 2007. A BBC television crew in Peru in 1998 filmed Blashford-Snell's fiasco on the Desaguadero, along with serious discussion of pre-Columbian voyaging between Polynesia and South America—including me on a Pacific beach saying it was probable although infrequent—and this film was later televised in a BBC series called "Supernatural Science." To my dismay, the film is titled "Atlantis Found!" Don't believe it.

■ ■ ■ ARYANS

Atlantis out beyond the Straits of Gibraltar is a harmless parable. Not so the Aryans, another fantasy about the past that was foisted on a willing public. Hitler admired the "Aryan race," said to be big, blond, blue-eyed masterful men and strong women. Hitler himself was short, dark-haired, and brown-eyed. That didn't deter him from promoting "Aryan types" to leadership in the German military and bureaucracy and creating "maternity homes" where "Aryan"-looking young women were kept to become pregnant by "Aryan" army officers. Non-"Aryan" Jews, Gypsies, and eastern Europeans were, as you know, gassed by the millions.[6]

Under Hitler's rule, German archaeologists were expected to expose sites demonstrating the invasion and settlement of Germany by the Aryan master race. The scenario, formulated in the nineteenth century, had thundering chariots driven by blond hunks, spear in hand, hurtling out of the steppes onto degenerate dark-haired people's cities. Pure-souled "Aryans" regenerated through conquests, establishing a higher civilization and nobler religion. This was Adolf Hitler's mission. Needless to say, during der Führer's deadly decade of power, archaeologists who wished to remain in Germany did not dispute his version of history.

[6] See essay by Leach on unjustified projections of an "Aryan race." Bettina Arnold's 1990 article describes Nazi archaeology.

The conviction of "Aryan superiority" has never completely disappeared. It appears in some American groups battling for "White supremacy"—that is, privileged status for them and denial of human rights to persons with non-European ancestors. When James Chatters, the archaeologist who analyzed Kennewick Man found beside the Columbia River in Washington state, mentioned that the 9,400-year-old skeleton looked "Caucasoid"—meaning like older Eurasian humans rather than like more recent American Indians—an "Aryan" group demanded custody of the skeleton. No archaeological or biological data have substantiated the supposed "Aryan race," but the label lives on.

Though lacking the racial overtones, the alleged existence of Aryan peoples has also colored interpretations of European prehistory by seeing England, for example, advanced toward civilization by a series of invasions from continental Europe. One British archaeologist argued as late as the 1980s that Indo-European speakers—"Aryans" were said to speak an Indo-European language—introduced farming to Europe by invading out of Turkey. (Current opinion sees farming spreading into Europe from the eastern Mediterranean several thousand years earlier than Indo-European languages; see Chapter 9.) The appearance of new artifact types, especially weapons and apparent luxury goods, could be attributed to invading hordes rather than undramatic merchants' business enterprise. Again and again, archaeological data, from "kurgan" burial mounds to battle axes to the ruins of Indus cities, have been shown to be equally plausibly interpreted without invoking legendary Aryans.

■ ■ ■ Dinosaurs

As his notorious son Jeffrey Dahmer murdered half a dozen men and boys trying to turn them into his zombie slaves, industrial chemist Lionel Dahmer worked on an award-winning paper proving humans and dinosaurs had lived at the same time, around thirty thousand years ago. This was not simple ignorance that dinosaurs became extinct millions of years before even the earliest hominids evolved; this was an effort to disprove evolution. Dahmer's award came from an organization of fundamentalist Christians rejecting any "theory" at odds with their selective, literal reading of Genesis in the Bible. They deny the earth began more than a few thousand years ago, and they insist that every kind of organism was made by God as we see it today. Since they invoke a supernatural deity, their beliefs are outside the domain of science (whether or not they might be true). Nonetheless, they want the world to accept their views as legitimate. And because our society looks to science for truths, they present their ideas as science—"Creation Science."

Lionel Dahmer and friends went to their nearest museum, Carnegie Museum of Natural History in Pittsburgh, and asked the curator of paleontology for a few chips of dinosaur fossil they needed for an experiment. He took them to a collections drawer, showed them some chips from a fossil, and let them take them. They sent the chips, with a check for the processing fee, to a radiocarbon dating laboratory. The lab wrote back offering to return the sample and fee because the chips did not contain collagen (the protein that contains carbon in organisms), and anyway, analysis would be affected by shellac contaminating the sample. The submitters wrote back, insisting that the lab process their sample. The lab report arrived: the machine gave estimates on the chips of > 23,760 and > 25,760 ± 270 years. YES! The non-believers' own scientific laboratory confirmed that dinosaurs lived during the same time they assigned to early humans! For this triumph, Dahmer won the Creation Research award.

Okay, readers, what did Lionel Dahmer overlook? The radiocarbon lab told his group that their sample lacked the organic matter that was normally radiocarbon-dated. Competent archaeologists understand that the air we and other organisms breathe contains carbon dioxide in two forms, carbon-13 and its radioactive isotope, carbon-14. When an organism dies, the radioactive carbon-14 decays at a regular rate, half of it being gone after a bit more than five thousand years. By finding how much radioactive carbon is left, a laboratory can estimate how long ago the organism died. Fossils, however, are inorganic *mineral replacements* filling space left by decaying organic bone; there is no organic carbon in fossils, so they cannot be carbon-dated. The small amount of carbon used to date the fossil chip was an inorganic carbonate having nothing to do with any living creatures. Dahmer didn't understand that the fossil "bone" was like a mold from the completely decayed original bone. Though trained in science, his lack of knowledge of the basics of archaeology led him (quite willingly) to a completely erroneous conclusion, proving to his supporters something that has been clearly invalidated by over a hundred years of research on dinosaurs.

More evidence that humans and dinosaurs roamed together is said to be on view along the Paluxy River in Texas. Dinosaur footprints are visible in a rock formation exposed along the stream. According to the Creation Research followers, human footprints, giant ones at that, appear alongside the dino prints. Paleontologists examining these prints concluded that the "human" prints are dinosaur tracks, only less well preserved and therefore distorted. The Creation Science advocates insist they see human toe marks in the rock, and advertise their museum showcasing their story of God creating dinosaurs along with humans. (Apparently, dinosaurs were too big to go into Noah's Ark, so they perished in the Great Flood.) Creation Science believers also buy figurines of men riding dinosaurs, giant apes, and extinct mammoths. These "proofs" of humans and dinosaurs together are made by farmers, some in Mexico and some in Peru, to sell to credulous local collectors and to tourists. Although the Peruvian farmer who first, in 1966, sold carved stones depicting dinosaurs and humans was arrested for fraud, the tourist market for fakes keeps the business going. Anti-evolution creationists and gullible or amused travelers don't consult paleontologists before handing over money.

Dinosaurs, who went extinct 65 million years ago, and humans, whose modern species came into being only about 150,000 years ago, didn't coexist, no matter how much Creation Scientists want you to believe it.[7]

■ ■ ■ POPULAR ARCHAEOLOGY CONTROVERSIES

The television films that I participated in, "In Search of Atlantis" and "Atlantis Found!," expose how the public is led to think that "controversies" are so common in archaeology: television shows and books and magazine articles aimed at the buying public use drama to sell. "In Search of Atlantis" was carefully researched and presented highly reputable archaeologists to explain to viewers what is known about Greece in the Homeric Age, about the earliest civilization in Mexico, and about a South American empire preceding the Inca. Any controversies about these sites are on the level of details discussed among specialists, not about basic interpretations heard in the film and definitely not about whether any could have been Plato's Atlantis. There are probably nothing near fourteen million people who *care* about Atlantis, but there are millions willing to be intrigued about esoteric topics hyped as "secret knowledge."

Part of the set-up lies deep within Western culture. Scholars studying the Indo-European languages from which Western culture developed, and mythologies associated with them, discern a worldview supposing conflict to be frequent and inevitable.[8] St. Augustine taught that to make peace, one has to first make war!—excusable, perhaps, in a man destined to be killed in the siege of his city, Hippo in North Africa, though hardly compatible with professing allegiance to the pacifist beliefs of Jesus of Nazareth (who spoke a Semitic language, not Indo-European). American society follows this Indo-European worldview and

---

[7] Access the National Center for Science Education web site to see anti-evolution claims and scientists' responses: www.ncseweb.org

[8] See my essay, "Conflict is a Western Worldview."

speaks an Indo-European language as well, encouraging competition and ranking. Anyone who looks coolly with an anthropologist's eye can see that most of the time, most humans are inclined to cooperate, not compete—that is why we *teach* competition—and cooperation is obviously how we have survived over several million years, most of us living peacefully with others. People get excited by conflict, adrenaline flows, exactly because it is *not* our everyday state. People get excited by hyped conflict in film or books, and then feel good, relieved, when the scenario's conflict is resolved. People anticipate the pleasurable end after the adrenaline surge, and tune in or buy the story. It would be nice if people preferred the quieter pleasure of intellectual discussion, what one usually sees when archaeologists get together, but it's not considered red-blooded American: on Monday nights everyone is expected to watch football, not "NOVA" documentaries.

And let us repeat: Archaeologists do *not* dig dinosaurs. Dinosaurs went extinct millions of years before humans developed. Popular "knowledge" is like a scrapbook: a snippet here, a picture there, and not much effort to construct a logical, ordered exposition. Controversies that most people associate with archaeology are generally *not* controversial among archaeologists, or, as with dinosaurs, not even archaeology. This chapter has reviewed a variety of stories that we archaeologists hear again and again. We could call them contemporary folklore: they sell books and fill television hours. In the next chapters, I'll discuss topics that really are controversial among archaeologists.

SOURCES

Arnold, Bettina, 1990. The Past as Propaganda: Totalitarian Archaeology in Nazi Germany. *Antiquity* 64:464–478.

—. 1998. The Power of the Past: Nationalism and Archaeology in 20th Century Germany. *Archaeologia Polonia* 35–36:237–253.

Bahn, Paul G., 1996. *Cambridge Illustrated History of Archaeology.* Cambridge: Cambridge University Press.

Bernal, Martin, 1991. *Black Athena: The Afroasiatic Roots of Classical Civilization*, vol. II. New Brunswick, NJ: Rutgers University Press.

Evans, Susan Toby, 2004. *Ancient Mexico and Central America: Archaeology and Culture History.* London: Thames and Hudson.

Hamdani, Abbas, 2006. Arabic Sources for the Pre-Columbian Voyages of Discovery. *Maghreb Review* 31(3–4):203–221.

Hornung, Erik, 2001. *The Secret Lore of Eygpt: Its Impact on the West.* Translated by David Lorton. Ithaca, NY: Cornell University Press. (Original: *Das esoterische Aegypten*, 1999, Munich: C. H. Beck'sche.)

—. 1999. *The History of Ancient Egypt: An Introduction.* Ithaca, NY: Cornell University Press.

Kehoe, Alice B., 1992. Conflict is a Western Worldview. In *The Anthropology of Peace,* edited by Vivian J. Rohrl, M. E. R. Nicholson, and Mario D. Zamora, pp. 55-65. Studies in Third World Societies, no. 47. Williamsburg, VA: College of William and Mary, Department of Anthropology. (Reprinted 2000, in *Social Justice: Anthropology, Peace and Human Rights* 1 (1-4):55-61, edited by Robert A. Rubinstein. IUAES Commission on Peace & Human Rights. Syracuse, NY: Syracuse University Maxwell School.

Leach, Edmund, 1990. Aryan Invasions over Four Millennia. In *Culture through Time: Anthropological Approaches,* edited by Emiko Ohnuki-Tierney, pp. 227–245. Stanford, CA: Stanford University Press..

Lepper, Bradley T., 1992. Radiocarbon Dates for Dinosaur Bones? A Critical Look at Recent Creationist Claims. *Creation/Evolution* 12(1):1–9.

Mayor, Adrienne. 2005. *Fossil Legends of the First Americans.* Princeton: Princeton University Press.

Pauketat, Timothy R. 2004. *Ancient Cahokia and the Mississippians.* Cambridge: Cambridge University Press.

—. 2008. *Cahokia's Big Bang and the Story of Ancient North America.* New York: Viking/Penguin.

Preston, Douglas, 1995. The Mystery of Sandia Cave. *The New Yorker,* vol. 71 (June 12, 1995):66–83.

Schwartz, Stephan A., 1978. *The Secret Vaults of Time: Psychic Archaeology and the Quest for Man's Beginnings.* New York: Grosset and Dunlap.

Stafford, Thomas W., 1992. Radiocarbon Dating Dinosaur Bones: More Pseudoscience from Creationists. *Creation/Evolution* 12(1):10–17.

4 America's First Nations and Archaeology

The most serious legitimate controversy in contemporary archaeology is the question of whose country America is. America is called a settler society, inhabited by Europeans only since the fifteenth century, as are Australia, South Africa, and New Zealand. The precolonial pasts of these nations, and others colonized by European powers, are usually excavated by archaeologists trained and employed by descendants of the imperial nations that invaded and eventually dominated others' homelands. Do remains of these pasts belong to the conquerors or to descendants of the indigenous peoples?

Controversies arise on several levels: Should excavations be directed by outsiders or by members of the indigenous group? Should finds be taken away, or housed in the homeland? Should outsiders' interpretations prevail over indigenous traditions? Should archaeology be conducted if indigenous people believe it would offend their values? At one point, these were not issues. The colonial powers, and the archaeologists who worked for them, believed they knew best how to govern and how to understand the past. But the peoples of the less developed world who became independent in the late twentieth century and the indigenous inhabitants of settler societies began demanding rights over their heritage. Complicating the issues is the political strategy of enlisting support by playing upon sentiments such as letting the dead rest in peace. A First Nation—that is, one of the nations first in a territory, before European invasions—may not be unified, pro or con, and who should speak for the nation may be contested.

Indigenous nations were generally ignored by archaeologists until late in the twentieth century. Conquering them had been legitimated by claiming that they were savages or infidels destined to be overcome by Christian gentlemen (see

Universes of Discourse

Everyone who has tried to translate from one language to another knows that one can't just substitute words from parallel word lists. "Boy" in English can be *niño* or *muchacho* in Spanish: which do you use? The archaeologist called in to identify the skeleton found eroding out of the Columbia River bank remarked that he looked "Paleo-European," a technical term referring to northern Eurasians of around ten thousand years ago. Journalists were confused; they assumed that meant he had been of European descent, and a small sect of White so-called Neo-Pagans wanted to rebury him with their reconstituted pre-Christian rites. Archaeologists are familiar with the considerable differences that develop within populations through the four hundred generations that occur over ten thousand years. Scientists live in mental universes composed not only of technical terms but also of concepts they consider obvious or well established. Such a "universe of discourse" marks persons trained in a scientific field, as other sets of terms and concepts form universes of discourse for sports fans, lawyers, fashion designers, bridge players, cooks, college students—we all "code-switch" talking to parents, to our buddies, to co-workers. People brought up speaking a particular language, in a family or community sharing that language, religion, music, cuisine, and expectations of social behavior, live in a universe of discourse that may seem radically different, stupid, or wrongheaded to someone from another cultural background, accustomed to another universe of discourse. The same goes for scientists in their universes of scientific discourse, versus people without such training: science by definition deals only with the actual world, excluding discourse about supernatural beings and events. People whose social universe includes talking about God sometimes denounce scientists as godforsaken. Archaeologists may find themselves inveighed against by Christians who insist their edition of the Bible is absolutely true, and also by psychics, Goddess-worshippers, and followers of non-Western religions, all universes of discourse admitting concepts a scientist, *acting as a scientist,* cannot accept.

Chapter 10, "Delgamuukw"). Remnants of First Nations were considered either still dumb savages, or to have lost traditional knowledge under forced Western schooling. Archaeologists talked of using "ethnographic analogy"—descriptions of non-Western communities—to interpret precolonial sites, but few sought first-hand experience with non-Western ways of life. Most simply read social anthropologists' accounts of "tribal peoples" in British, French, or German colonies, or descriptions of American First Nations collected in the years after they were forced onto reservations. Generalizing from such accounts and projecting the generalizations back into times before European overseas expansion tended to perpetuate stereotypes. Women, although until recently a minority among archaeologists, have been more likely to have conducted ethnography as well as archaeology. Probably this is due to girls being socialized to be people-oriented, and boys socialized to be more confrontational and attracted to tools and technology.

A notable example of an archaeologist committed to working with First Nations people in her research area was an American, Frederica de Laguna (1906–2004), who collaborated with Danish colleague Kaj Birket-Smith in Alaska. (Reminiscing decades later, de Laguna remarked that Danish men were more comfortable collaborating with female colleagues than were Americans. She had previously traveled 1,600 miles down Yukon rivers in a pair of skiffs with two American geologists and a linguist, all male, she looking for Paleoindian sites, and that was pioneering in more ways than one.) She and Birket-Smith took turns alternating between supervising field excavations, and ethnographic observation and interviewing in the local indigenous community. Community members made up the project's work crew and participated in creating the interpretations. Years later, grandchildren of the participants invited de Laguna to be honored with a feast in the community for her substantial contributions to their history.

The Smithsonian Institution published de Laguna's reports in detail. Look at the long list of her monographs and books, and you'll see two unusual titles: *The Arrow Points to Murder* and *Fog on the Mountain*. These are fast-paced murder mysteries drawing on her professional experience, the first taking place in a major museum and the second in southwestern Alaska. The latter, describing the village in Prince William Sound where she and Birket-Smith worked, is relevant for our topic of controversies in archaeology. In the story, there is only the young man ethnographer, clearly Birket-Smith, alone in the remote locality where the Indian people reside separately from the small White community of storekeeper, constable, fish cannery manager, and commercial fox raisers (for fur coats). Our hero is recording the native culture from the elderly chief, interpreted by his daughter Matrona, when the old man is killed. Dastardly deeds are done, rumors abound, our hero comes perilously close to losing his life, saved by remembering a legendary trick in a traditional tale. Happily for Dr. Birket-Smith, that's all fiction, but the details of village life in the 1930s are true, down to the chief's daughter's name: the real Matrona Tiedemann interpreted for her father, the chief (Figure 4.1).

FIGURE 4.1 Frederica de Laguna, *center*, with Danish anthropologist Kaj Birket-Smith, *right*, and Chugach collaborators Matrona Tiedemann, *left*, and her father Chief Makari. The man at the far right is unidentified. *Photo from Florence Hawley Ellis Archives, Albuquerque.*

De Laguna always sought, as she put it in the title of one of her Smithsonian monographs, "the relationship between archaeological, ethnological, and historical methods." She noted in that study that the archaeologist's collections will tend to represent the "junk" of everyday life, since

> he [even de Laguna wrote "he"] finds chiefly what people have lost or thrown away, whereas the museum's ethnological collection will more likely contain a greater proportion of handsome "exhibits." . . . The same sort of comparison can be made between the written accounts of early visitors and the monograph of the ethnologist who records the oral reports of a vanished way of life. (de Laguna 1960:2–3)

Early visitors either didn't stay long in a community or had jobs to do that limited their experience of indigenous life, in contrast to ethnographers (ethnologists) immersed full-time listening to and observing native people. De Laguna constantly asked Tlingit people, in whose homeland she was excavating, about artifacts she uncovered, and encountered quite uneven knowledge:

> [S]tone adzes, harpoon heads, beads, pendants, and labrets [ornaments inserted under pierced lips or beside mouths] were objects everyone recognized. . . . The needle, however, has so completely replaced the [bone] awl that no one seems to have any clear ideas of what a bone awl was really like . . . descriptions of old-style houses were often vague and stereotyped, for the modern framehouse replaced the aboriginal plank house when today's [1950] old people were children. Yet the smokehouses still in use represent the whole series of types from the oldest to the most modern. (de Laguna 1960:15)

Another source of "discrepancies" is differences between societies regarding the way they talk about their land. De Laguna and Birket-Smith's Chugach

Yu'pik native teachers at Prince William Sound recounted histories of famous leaders' journeys, specifying exactly where events took place, and the two anthropologists found archaeological evidence at many of the sites. Tlingits living south of the Chugach were less helpful because most knew in detail only their own lineage's history places, obliging the anthropologists to talk with more individuals and figure out overlapping mappings. Frederica de Laguna's decades of intensive archaeological and simultaneous ethnographic and historical research in Alaska[1] and the Yukon demonstrate how intelligent archaeologists can enrich interpretations of material culture, not just by reading ethnographies but by spending field time talking at length, and respectfully (she emphasizes that), with local indigenous people. Her work has become a model for contemporary collaborative archaeological work with indigenous communities.

■ ■ ■ WHO OWNS INDIGENOUS HERITAGE?

Epidemic diseases, a century or more of warfare, and post-conquest poverty decimated First Nations after Columbus's 1492 landing. First Nations were banished to reservations often thousands of miles from their homeland, their indigenous languages were outlawed, and their children were forcibly removed to attend Western schools. During the twentieth century, their populations began recovering. Civil rights edicts and human-rights campaigns beginning in the 1960s coincided with increasing First Nations numbers and participation in mainstream society, leading to legislation reducing domination by federal bureaucracies. The Anglo-American countries, the United States and Canada, had originally concluded treaties with most of their aboriginal nations on a basis of recognizing indigenous sovereignties. Legally, in spite of U.S. Chief Justice John Marshall's 1831 odd ruling that American First Nations had become "domestic dependent nations," by treaty they retained sovereignty except where explicitly delegated to the Anglo power. Canadian First Nations took advantage of the repatriation of Canada's constitution from Great Britain, in 1982, to demand to be included with premiers of the provinces in meetings with the federal government, the elected Chief of the Assembly of First Nations to be seated as representative of that component of Canada. Indian nations in the United States were slower to demand greater recognition of sovereignty, worrying that it might reduce economic assistance. After a generation of testing the waters, as it were, U.S. "tribes"

[1] De Laguna's work in the North was interrupted by World War II, when she served in Naval Intelligence. Many archaeologists spent those war years in military service or related duties.

(a denigratory term still used in the United States) shifted to asserting their rights both under treaty and as human rights. The watershed was 1990, when Congress passed NAGPRA, the Native American Graves Protection and Repatriation Act.

For Congress, NAGPRA was feel-good legislation, protecting graves from desecration. In addition, it provided for repatriation (return) of skeletons and associated grave goods to the bodies' kin, if that could be determined, and of sacred objects to their original tribes. The congressional bill was poorly written, leaving some critical terms ill defined. It mandated federal institutions, and museums and universities receiving federal government funding (even if very limited), to list all aboriginal American human remains, artifacts associated with them, and other indigenous heritage artifacts in their collections, to make the lists available to the kin or tribe to which the graves belonged, and if requested by kin or tribe, return ("repatriate") the remains and artifacts. In effect, America's First Nations people were now to be respected rather than held to be of no account. But Congress did not mandate money for the daunting task of working item by item through museum catalogs, checking each entry against the actual object in storage, sorting out lists by tribes, and contacting the tribes. Nor did NAGPRA distinguish between direct kin of a burial, whom no one would deny the right to receive the remains, and ancient humans buried in a locality that centuries or millennia later became the territory of a historic nation.

This oversight led to the notorious case of Kennewick Man, whose skeleton eroded out of the bank of the Columbia River near the Umatilla Indian Reservation in central Washington state (Figure 4.2). Radiocarbon dating demonstrated that Kennewick Man lived 9,400 years ago, he had been wounded with, but survived, a stone spear blade still embedded in his hip, and physically he looked more like people of northern Asia of that time period ("Caucasian," as we mentioned in the last chapter) than like today's Umatilla. Nevertheless, the Umatilla and Colville Reservation tribes (ten groups had been forced onto Colville in the nineteenth century) forbade archaeologists and biological anthropologists to examine the skeleton they named "the Ancient One" and demanded it be given to them to be reburied with their ceremonies. Such a demand is by no means universal among Indian people, many of whom are disturbed that such ancient remains, or any that aren't historically known to have been members of a particular First Nation, may be arbitrarily subjected to the religious rituals of what likely was a foreign group, perhaps even an enemy. Many Indian people want scientists to find out all they can about ancestral populations, knowledge that need not necessarily conflict with religious traditions about ultimate origins. After ten years of bitter court battles, Kennewick Man was released to the coalition of anthropologists who had sued to analyze it, by a

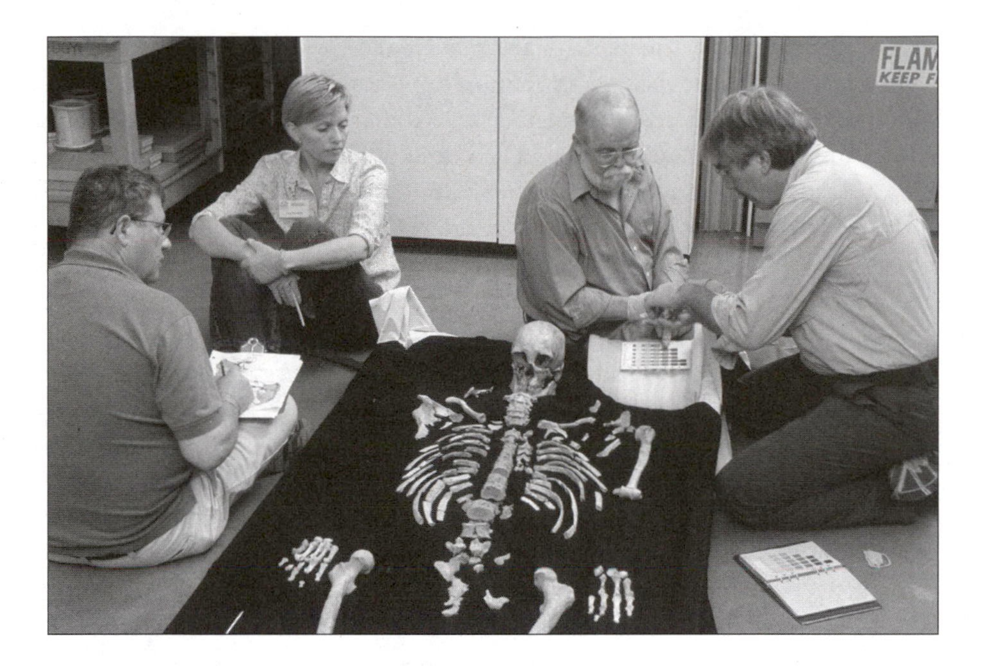

FIGURE 4.2 Kennewick Man, the 9,400-year-old skeleton from western Washington state. After a bitter legal battle, the court permitted scientists to analyze the bones. The case clarified the law designed to return excavated human remains to their tribes. We can't determine which First Nations today (if any) might be related to such ancient Paleoindian skeletons. *Photo by Chip Clark.*

federal judge in recognition of the nine-thousand-year gap between the skeleton and historic American Indian communities along the Columbia River. Overall, the Kennewick case is an exception to general goodwill between First Nations and museum professionals working hard to comply with NAGPRA and build constructive collaborations.

■ ■ ■ COLLABORATION

Quite the opposite from the Kennewick Man case can be seen in the cooperative handling of Kayasochi Kikawenow, a Cree Indian woman who lived around A.D. 1665 in northern Manitoba, Canada (Figure 4.3). Local men noticed her burial eroding from the shoreline of Southern Indian Lake in 1993, similar to the way Kennewick man would be discovered along the Columbia River three years later. The archaeologist supervising research in the region was called in, and with the assistance of a contract archaeologist and a crew from the community, excavated and recorded the burial, struggling against the lake water seeping around their makeshift coffer dam. The remains and grave goods were sent to the Manitoba Museum of Man and Nature in Winnipeg, where a Cree graduate student archaeologist and a variety of scientists analyzed the burial under the direction of senior anthropologist Leigh Syms. Their results were shared with the South Indian Lake and Nisichawayasihk First Nation Cree communities, and when the analyses were completed and replicas made of the artifacts, Kayasochi Kikawenow was placed in a small wooden coffin with sweetgrass, sage, tobacco, and her original grave goods. The young Cree archaeologist, Kevin Brownlee, carried her back to her home and dug a new grave for her safe from the lake waters. A Cree minister serving the local church conducted a funeral ceremony.

Kayasochi Kikawenow, "our Mother from Long Ago" as the Crees called her, taught her latter-day people a great deal about their forebears. The contemporary elders believed she had manifested herself in order to pass on knowledge of her time and reinforce the community's commitment to their heritage; archaeologists and laboratory scientists were instruments for her purpose. Brownlee prepared a permanent exhibit on Kayasochi Kikawenow for the village school, along with materials for teachers, while the Manitoba Museum published the well-illustrated report he and Syms prepared (Brownlee and Syms 1999). These records are the gift of the Crees' *kayasochi kohkominow* "long-ago grandmother," gratefully acknowledged at a community feast of moose meat and moss-berries, bannock and tea (Kayasochi Kikawenow, in 1665, saw a few European imports passed on through Indian traders, perhaps including tea). Cree and scientific universes are compatible when each understands and respects the differences in the domains acknowledged in their universes of discourse.

Kevin Brownlee is one of an increasing number of archaeologists who are members of American First Nations. This generation is by no means the first: a century ago, Dr. Arthur Parker, a Seneca Iroquois, was New York State Archaeologist and first president of the Society for American Archaeology, and a Dakota, Amos Oneroad, assisted Alanson Skinner excavating a number of prehistoric sites in, of all places, New York City (Wall and Cantwell 2004:79).[2] Best known among contemporary American Indian archaeologists is Joe Watkins, a Choctaw who worked as a contract archaeologist and for the Bureau of Indian Affairs before accepting a faculty position at the University of New Mexico. Watkins's book, *Indigenous Archaeology: American Indian Values and Scientific Practice*, published in 2000, candidly describes his efforts to counter both Indian people's accusations that all archaeologists are nothing but grave-robbers, and non-Indian

[2] Wall and Cantwell assume Oneroad was Menomini because Skinner worked with that nation. Robert L. Hall, who has Menomini relatives, informs me that Oneroad was Dakota.

FIGURE 4.3 "Kayasochi Kikawenow," a Cree Indian woman in Manitoba, Canada, who died about A.D. 1665. It is reasonable to link her to Cree people in the vicinity three centuries later. This Indian community invited archaeologists and other scientists to analyze the remains, believing that the woman's spirit revealed her in the eroding grave so that her descendants can know more about their forebears. *Drawing by Don McMaster, The Manitoba Museum, Winnipeg MB.*

archaeologists' tendency to consider their science-based interpretations the truth, more valid than First Nations' orally transmitted histories.

Many First Nations now employ archaeologists to advise in protecting their heritage, to survey and to test ancestral sites. At times, professional archaeologists (whether or not they are themselves American Indian) may be uncomfortable with tribal employers' plans, a situation that usually stimulates discussion to explore sources of disagreement. The question of whether Indian communities and archaeologists can collaborate is resoundingly answered positively, as every year more First Nations hire more archaeologists (Figure 4.4).

Hopi Archaeology

Northern Arizona is home to Hopi, Zuni, and Navajo Nations. Each maintains an archaeological program to identify, protect, or if unavoidable, salvage sites, document territorial claims, and complement histories traditionally passed down orally. Although Navajo (Diné) moved into the American Southwest only in the fifteenth century and filled in open lands between the permanent towns of Hopi and Zuni pueblos, all three nations' histories tell of migrations. Archaeological research to detect clan migrations has been encouraged by the

FIGURE 4.4 Hopi middle school students in the Hopi Footprint Project work with professional archaeologists at an ancestral Hopi village in Homol'ovi State Park. Many Indian communities are inviting archaeologists to involve their young people in projects on ancestral land. *Courtesy Hopi Cultural Preservation Office and Northern Arizona University.*

Hopi Cultural Preservation Office, bringing together archaeologists, professional linguists and their Hopi consultants, historians, cultural anthropologists, and Hopi civic and religious leaders. Roles overlap, in that some of the academically trained personnel are themselves Hopi.

Old-fashioned anthropology made the assumption that small agricultural nations lived generation after generation in their valleys, seldom venturing beyond surrounding hills, passing down technologies and artistic styles that changed slowly and without conscious innovation. Hopi historians, in contrast, recounted instance after instance of communities ("clans") trekking from distant locales, past series of landmarks, to the Hopi mesas where earlier settlers allowed them to take up land and add their rituals to the hosts' rich store. Conversely, it is told that certain groups separated from mesa villages after political crises. Hopi "influence" in the form of distinctive pottery styles was noted ever since archaeologists began working in the Southwest in the late nineteenth century, and was assumed to represent trade, perhaps village to village or else by a few itinerant traders. In the 1990s, encouraged by the Hopi Cultural Preservation Office, projects were initiated to determine, by laboratory tests of chemical elements in clay, whether "trade" pottery had been manufactured on the Hopi home mesas and carried away, or made in other regions by potters who may have moved from Hopi lands or learned to work in the Hopi style. It turned out that both scenarios happened. More significantly to the Hopi, places that their historians had claimed were sources or endings of migrations were validated when Hopi ceramic styles were found to be manufactured in those places.

As archaeologists and Hopi staff of the Hopi Cultural Preservation Office discussed their various data, what at first appeared to be conflicts opened up new understandings. Hopi realized that instead of assuming there is nothing more to discuss than a single place their remote ancestors emerged from—maybe in the Grand Canyon west of the Hopi mesas—they could envision an ancient *time* during which their ancestors developed Hopi culture, building and moving settlements within most of Arizona. Archaeologists realized that each Hopi clan was telling its particular history; these are not a national history for today's Hopi Nation, but multiple strands interwoven and sometimes unraveled. We could remark, Whaddaya know! Hopi have a real-world history not unlike that of other nations, people moving from here to there and maybe back again, or maybe farther on, pulled by prospects of better resources or pushed by enemies, droughts, floods, or quarrels. Hopi, living on three large mesas when the United States took over Arizona in the mid-nineteenth century, represent an aggregation of histories and a way of life constructed through accommodations, compromises, and creative breakthroughs. Grasping this point of view, archaeologists have charted movements of proto-Hopi in the thirteenth through sixteenth centuries from the Four Corners area of northernmost Arizona, into all the river valleys of the southern half of the state, and occasionally back north. Fine-tuned archaeology and oral histories agree that in most cases, Pueblo settlements were ethnically diverse, harboring refugees and immigrants offering desired skills, labor, spiritual knowledge, or military assistance. This newer understanding shows that prehistoric Pueblos were much like those of the historic period, not the peaceful isolated little communes pictured by earlier archaeologists.

Building on this approach, cultural resource management archaeologists in Arizona developed a project to explore cultural ties to one valley, the San Pedro River Valley in southeastern Arizona. (Cultural resource management, usually called CRM, refers to the practice of discovering and protecting heritage properties such as archaeological sites or historic buildings. Due to federal, state, and municipal laws guarding against wanton destruction, particularly since the 1970s, thousands of archaeologists are employed in finding, safeguarding, or rescuing both prehistoric and historic sites.) Four First Nations with oral histories tied to the valley were invited to participate: O'odham, Apache, Zuni, and Hopi. O'odham and Apache occupied the valley in the nineteenth century, and the two Pueblo nations, Zuni and Hopi, presently a couple hundred miles (about 260 kilometers) north, recount both migrations to and from the valley, and trade with others in it. Each of these nations has its own program for historic and cultural preservation, with which the San Pedro Valley project could cooperate. It happened that the project researchers are non-Indian, and they noticed that during many hours of talking with representatives of the four Indian nations, much of the time out in the field, their Indian collaborators were doing ethnography on them, discovering their "Anglo" values and ways of thinking. Frederica de Laguna mentioned that she and her team members were often the first White people to listen at length and respectfully to the Indian people they worked with. I'm sure some of my own Indian teachers were observing American culture through me. Anthropologists may be the only non-Indians whom some older people living in Indian communities have had the opportunity to really get to know.

The San Pedro River Valley study illuminates some of the complexities underlying historic archaeology from a First Nations standpoint, and the problems NAGPRA produces. In the hot, dry, southeastern corner of Arizona, the valley is within a "desert culture" area described from late nineteenth-century ethnographies. O'odham (formerly called Pima and Papago) are the principal indigenous groups, with Western Apache added as protohistoric migrants. O'odham speak Uto-Aztecan languages related to those of northwestern Mexico, the

Aztecs, and Hopi Pueblo. It was reasonable for archaeologists to label most of the ruins in southern Arizona "Hohokam," to surmise that forebears of the O'odham made them and that Hohokam culture was linked to, and may have come from, northwestern Mexico. O'odham themselves say that they invaded and conquered an earlier people who built the big ruins such as Pueblo Grande in today's city of Phoenix.

Zuni, who speak a language apparently unrelated to any other existing language, and Hopi say that some of their ancestors moved to southern Arizona, a tradition supported by Western Pueblo Late Prehistoric pottery and artifact styles in pueblo ruins in the San Pedro Valley. When the Spanish invaded from Mexico in the early sixteenth century, no one in the Valley was living in big adobe buildings; instead, O'odham were scattered in hamlets of small brush or adobe homes, their material culture quite distinct from either Hohokam or Western Pueblo. No one knows whether those societies abandoned more complex organizations and elaborate material culture to live a simpler life closer to sustainable subsistence in the desert region, whether the Western Pueblo groups moved back north to their relatives, whether ethnic Hohokam fled somewhere where they could no longer carry on their culture, or morphed into O'odham. Are these questions controversial? Not in terms of land claims, which require historic documentation (of the time period in which the United States and Canada existed), but yes in terms of NAGPRA assignments for obtaining tribal permission for excavations and repatriating human remains, grave goods, and sacred objects to their descendants. Studies such as the San Pedro River Valley project, which was funded by the National Endowment for the Humanities, clarify grounds for identifying tribal affiliation, and set up relationships to call upon for future negotiations.

■ ■ ■ LAND CLAIMS

Among women archaeologists who worked also as ethnographers was Florence Hawley Ellis (1906–1991). She assisted Pueblo communities in documenting the occupation of land and associated water rights, highly important in the Southwest, considering it an ethical obligation to use the archaeological and historical data she collected for the benefit of the First Nations on the land. The list of nations she helped is impressive: Acoma, Hopi, Jemez, Laguna, Nambe, Navajo, Pojoaque, San Ildefonso, Santa Ana, Taos, Tesuque, and Zia. Disdained by some of her peers because she never wrote grand theory, Hawley Ellis used her expertise in dendrochronology (tree-ring dating, useful for dating structural timber in pueblos) and chemical sourcing, and her abundant excavation data, to collaborate with Pueblo people to build their histories.

Where a land claim, or a water-rights claim, is tied to a settlement that left ruins, First Nation occupation may be contested but would seem to be a matter of tying the ruins to historic descendants, commonly through distinctive pottery styles. What about claims to land used for hunting, not settlement? Can there be physical evidence for regular but transitory use? The Montana Blackfeet claimed land in the Rocky Mountain foothills along the western border of their reservation. The heavily forested, hilly terrain is prime country for elk, a significant resource for meat and leather for these people even today, and certainly was so before they were forced to live on a reservation. Bison (buffalo) had been their main staple, driven into corrals built under steep bluff drop-offs or in ravines. But bison hides were too heavy for clothing, necessitating hunting elk and deer for pliable leather, and elk meat was, and is still, much enjoyed. A classic example of hunter-gatherer culture, the Blackfeet had a definite seasonal round, traveling regular routes and camping to harvest root vegetables, berries, and animal game. Even before European traders came into their country, they produced surplus meat to dry and exchange with town-dwelling agricultural nations along the Missouri River. Sites of bison corrals and big camps where thousands gathered for ceremonies and trade are known through archaeological surveys and excavations, as well as Blackfeet histories, but the forested foothills west of the bison pastures and main tipi villages were just as much part of the Blackfeet nation's territory as the plains and river valleys where sites are visible.

The Blackfeet Tribal Historic Preservation Office hired an archaeologist to go into the claimed elk-habitat area with Blackfoot who hunted there from boyhood, and who learned from their fathers and other elders where they were accustomed to make their brief camps. Led by these collaborators to the locations, the archaeologist could notice rocks making a rough campfire hearth, the few artifacts in the surrounding underbrush, and faint trails. Putting these data on a map, she demonstrated regular and extensive use of the foothills territory for subsistence, enabling the Tribe to proceed with its claim that the area had been improperly left out of the Blackfeet Reservation.

Can They Be Trusted?

Who is "they" in the heading above? Indians? Anglos? Indian people learned to mistrust Europeans and their immigrant descendants after innumerable tricky promises and unjust actions. Thomas Jefferson's policy, during his presidency (1800–1808), was to build trading posts where Indians could be induced to go into debt (that favorite American practice of buying on credit), then be forced to sell their land to pay up, and then move west. Before too long, Jefferson figured, American expansion would push First Nations into the Pacific Ocean. After the Civil War, General Grant was elected President and in 1873 instituted a "Peace Policy" toward Indians: shift postwar military power to defeat First Nations, put ex-army officers in charge of Indian reservations, and then we shall have peace. He also used federal money to pay Christian church missions to educate Indians—never mind it's prohibited by the Constitution. Next, the federal government did its best to destroy the First Nation societies on the remnant reservations by decreeing their lands, held by their communities rather than by individuals, must be broken up into private household allotments where Indians should farm under Anglo instruction. One of the Smithsonian anthropologists helped with the Allotment Act, naively believing that it would relieve Indians of reservation agents' tyrannical dictates. It didn't take long to see that corrupt or incompetent agents let Anglo ranchers lease the allotments at below-market fees, fees that might well never reach the impoverished Indian families pressured to lease their plots. During the later nineteenth century and well into the twentieth, well-meaning self-appointed "Friends of the Indian" were as convinced as Jefferson that First Nations were doomed; they, as well as government bureaucrats, assumed Indian children could not succeed academically, Indians could not manage ranches or businesses, their religions were simple-minded superstitions, and the only course was to take away the children, to teach them manual trades they could use to work for Whites. In a country with this mind-set, archaeologists tended to assume there was a complete break with the past when an Indian nation was conquered, that after compulsory schooling, contemporary Indian people would not know much about their forebears.

Archaeologists once had little compunction about digging up Indian graves and removing skeletons and grave goods to museums for popular public displays. In my own city, the Milwaukee Public Museum exhibited for many years a complete skeleton adorned with broad belts of shell beads, excavated from a Wisconsin mound. This "Indian Princess Burial" wasn't quite what she seemed: the true skeleton was too fragile to mount for exhibit, so a Euroamerican male skeleton was substituted in its place. Who his kin had been, no one knows or seems to care. Up until the mid-twentieth century, anthropologists tended to think they were doing a favor for First Nations by buying ritual objects for museums, the notion being that reservation Indians weren't caring well for them. Sometimes this was true, due to reservation poverty, and sometimes the objects were saved from being destroyed by zealous missionaries or their converts. In one case, the Omahas' holy "Venerable Man" wooden figure was entrusted to Francis La Flesche, an educated Omaha employed as an anthropologist by the Smithsonian, soon after the Allotment Act had passed. La Flesche carried this embodiment of Omaha nationhood to Harvard's Peabody Museum for safekeeping. Exactly a century later, in 1989, a non-Indian anthropologist intervened on behalf of Omaha friends to persuade the Peabody Museum to lend Venerable Man to his people on their Nebraska reservation. Venerable Man was honored there with traditional rituals, and then returned to Peabody for protection until the next year's ceremony. Passage of NAGPRA that year encouraged and facilitated more First Nations to retrieve holy icons for rituals, with others (beside the Omaha) deciding to let holding museums curate fragile objects between ceremonial performances. Ironically, it may be necessary for museums to train the First Nations original owners how to handle the objects safely, because museum conservators, by repeatedly coating perishable objects with pesticides, inadvertently left such objects covered with poisons. Only recently have museum curators learned to test artifacts before repatriating them and to work with local peoples to avoid dangerous toxicity.

It's not a straight black-and-white issue, from either side. Here in the twenty-first century, there still are archaeologists who can't believe oral histories can be valid. Such a person insists that no one could repeat long epics word for word, despite studies by folklorists analyzing illiterate bards' remarkably consistent performances of treasured tales, and recitations of nonwritten histories by appointed officers in non-European courts. Francis La Flesche took down thousands of pages of Omaha and Osage (a closely related nation) religious texts from ordained priests who used repetitive forms and rhythm to aid in precisely remembering words they believed to be so powerful that a mistake could kill. We literate people grew up writing down everything—to-do lists, appointments on calendars, our thoughts in diaries, thank-you notes, shopping lists—we seldom are expected even to memorize poems or songs as our grandparents did. We look up everything to check our memory. (Some of my houseguests express astonishment at how many books surround me. I wouldn't dare write a text like this one without piles of books at hand to check what I think I know.) Archaeologists who distrust nonliterate persons' ability to recall significant knowledge reasonably accurately are presuming that all humans have been brought up the same. It's like saying highly trained Chinese acrobats can't do what we see them perform, because you and I can't do the same.

On the other hand, there are First Nations people who don't mind telling the intruding Anglo a good story. More importantly, there are many First Nations knowledge-keepers who will not reveal holy knowledge. It may be because it could kill the untrained person who tries to say or write it. Or it may be forbidden to repeat the knowledge to foreigners or to persons of the opposite gender, or outside one's lineage. Or perhaps the archaeologist looks too much like those other nosy Anglos who interfere with fishing, hunting, child care, and ceremonies. I've been told several times, by First Nations collaborators, that they figured they could trust *me* because they saw me in jeans and t-shirts, and one or two kids hanging on—I was definitely not from an agency office! Those of us who work with First Nations take time to establish a clearly reciprocal relationship, fulfilling requests from them and sharing data and records. If we still are told that we should not hear or see, or photograph, certain of the nation's intellectual property, we respect that.

Basically, oral information from First Nations people has to be considered in the same manner as written documents. Like written documents, oral historical material originates from particular times and persons, and limitations imposed by these particulars must be considered. Like written documents, oral histories may be propaganda or efforts to whitewash or glorify events. Neither written documents nor oral ones should be uncritically taken at face value. Are there empirical observations fitting the oral descriptions? How about de Laguna finding sites where the Chugach Yu'pik elder said historical events had happened? Does there seem to be bias in the material—for example, Tlingit people telling de Laguna only the events in their own lineage's history, leaving out those belonging to other lineages in the village? What rhetorical forms mold the telling? Are histories told as heroes' quests, or to teach a moral, to legitimate a possibly unsavory deed, or to glorify aristocrats? If I am told that the great-great-uncle of the man I am talking with was hunting on Chief Mountain and stumbled onto Thunder's lodge there, it would be stupid of me to think, "Oh, that never happened. Thunderbirds aren't real." I'm told that there is archaeological material on Chief Mountain—people hunted and camped there and left offerings out of respect for Thunder, whose presence in summer is amply manifest as thunderclouds mass at the mountain. I've also been advised, through the story, that it would be appreciated if I didn't fool around the mountain, disrupting pilgrimages to its shrines.

Like historical documents, like archaeological artifacts, oral histories of First Nations and other indigenous peoples can be a valuable source of information for interpreting the past. But like the other two sources, the evidence must be carefully sifted and matched with other data to derive the best possible inference.

The Battle of the Little Bighorn

Thomas Jefferson deliberately lied in the Declaration of Independence. He wrote,

> The history of the present King of Great Britain is a history of repeated injuries and usurpations . . . . To prove this, let Facts be submitted to a candid world.

Jefferson then listed twenty-seven "Facts," the last in the list claiming,

> He [King of Great Britain] . . . has endeavoured to bring on the inhabitants of our frontiers, the merciless Indian Savages, whose known rule of warfare, is an undistinguished destruction, of all ages, sexes and conditions.

King George did not provoke Indians to attack colonists, who, after all, were his subjects; and the Indians on the colonies' frontiers were farmers whose attacks were usually acts of resistance to land takeovers, enslavement, and murders. Jefferson certainly knew that many Anglo women and children had been adopted by Indian captors, not destroyed, and it was common for Anglo men to marry into Indian communities. If Thomas Jefferson lied in the Declaration of Independence, what document can we trust? There is no one whose words are totally above suspicion, and very few whose words can be routinely dismissed as untrue.

Indians' veracity became an issue in 1876, half a century after Jefferson's death. A consequence of Jefferson's policy of driving Indian nations westward was crowding along the Indian side of the advancing American frontier. After the U.S. Civil War concluded in 1865, that frontier lay along the western plains where indigenous nations such as the Blackfoot and Crow, and west-shifted Midwesterners such as the Cheyenne and Dakota, hunted bison. The grasslands that nourished bison herds were coveted by American ranchers and by thousands of farmers lured by railroad agents' sales talks to bring homesteaders to the railroads' vast acreage. In 1874, a U.S. Army exploring expedition went into the Black Hills of western South Dakota, part of the Great Sioux Reservation guaranteed to the Lakota Sioux by a treaty six years previously. The expedition, led by a Civil War officer named George Custer, found evidence of gold in the Black Hills, prompting a flood of get-rich-quick Americans into the Lakota and Cheyennes' richest hunting lands, where there were elk as well as bison, and a mountain and cave shrines held to be holy by these Indian people. President Grant's "Peace Policy" authorized several contingents of the Army to maneuver on the Great Sioux Reservation to cut off "wandering" (hunting) Indians from the Black Hills, the Missouri River and its steamboat traffic, and transcontinental railroads under construction. First Nations reacted as you would expect, lodging complaints with government agents against violations of their treaties, and meeting with one another to discuss means of halting the invasions of their treaty territories.

In June 1876, three U.S. Army units deployed to southeastern Montana to push Indians away from those areas attracting thousands of entrepreneurs and colonists. One of the units, the 7th Cavalry Regiment, was commanded by General Custer. His regiment rode into the valley of the Little Bighorn River, south of the Missouri, where scouts had reported a large Indian camp. The camp was indeed large, an alliance of several thousand Lakota, Cheyenne, and other concerned Indian soldiers, accompanied by many of their families. The Indian army met their U.S. opponents, routed one section attacking them in the river valley, and pursued the main group riding with Custer, downriver to a hill where Custer decided to make a stand. Well-armed with good repeating rifles, the Indian cavalry fired volley after volley at Custer and his troops and then closed in on those still fighting, killing every one—210 men. Most of the other Army units, away from the fatal hill, were able to retreat.

The "Custer Massacre" shocked Americans, particularly because it occurred just as the United States was about to celebrate its centennial, July 4, 1876. Many citizens reading the Declaration of Independence in commemoration of its centennial noticed Jefferson's line about "the merciless Indian savages on our frontiers." Custer's Last Stand quickly entered American mythology, the heroic golden-haired young general and his American troops fighting to the death to protect America from those merciless savages. There were also sober inquiries into the fiasco, interrogating both the U.S. soldiers from the surviving units and many of the enemy combatants. Their recollections spoke of fright and disarray among Custer's outmanned troop on the hill. Some Americans, unwilling to see the Indians' grievances and military ability, wanted to blame General Custer for reckless bravado. For a century, debates inflamed the sad story. For a century, the story was told from only one point of view, that of the surviving other units of Custer's command.

Then came archaeology. Meticulous survey of the hill where the 7th Calvary died retrieved and plotted spent ammunition, other military objects, horse gear, and personal items. The dead had been hastily buried by other Army soldiers three days after the battle, but subsequently most of the remains, reduced to skeletons, were removed to a mass grave or, in the cases of officers, sent to their families. The archaeological project tested a number of the original hasty graves, identified by stone markers erected fourteen years afterward, and confirmed that the stones more or less marked where Custer's men fell. The real question was the conduct of the troops during the battle: did they fight bravely as a unit, obeying their commander, or did they panic? Were Custer's tactics the best possible under the circumstances of landscape and a large enemy force, or was he foolish? Indians' accounts described both bravery, by some men, and fear and breakdown by others. Archaeology answered the questions through charting trails of rifle cartridges. Cartridge cases and the marks on them from firing pins, and bullet extractors examined under the microscope, allowed researchers to identify individual rifles. Because most of the battle artifacts are metal, the archaeological crew used metal detectors to systematically locate artifacts, carefully uncovering each detected object by hand and recording its exact location in three dimensions. These maps revealed the movements of the men firing their weapons, or—equally informative—areas with few or no artifacts, either avoided by combatants or where men ran without shooting. The archaeologists concluded that the troopers' training broke under the unexpectedly heavy Indian onslaught: they failed to form a skirmish line. As the troopers moved and fired erratically, Indians closed in.

Indian accounts helped the archaeologists by providing details that couldn't be found in the ground (Figure 4.5). For example, several Indians who had fought in the river valley recalled that the troopers down by the river rode gray horses, indicating that Custer sent his E Company to hold back the Indians there; E Company rode grays, the other four companies rode brown horses. Indians described watching Custer's men dismount to take firing positions on the hill, and Indians rushing in to capture the valuable horses. Many horses were nevertheless killed in the shooting. For a good part of the relatively brief final battle (less than two hours on the hill), Indians told, they crept closer through brush and behind rocks, firing when they caught sight of a trooper or, as the fight wore on, just over bodies of horses on the hill, noticing that troopers crouched behind the dead horses. As the desperation of their position became obvious, about forty troopers ran down the hill to the wooded ravine, hoping to get away. Indians described killing these unfortunates. The Indian accounts, including a few from women watching from the river valley, seemed confused to historians accustomed to reading official documents, because each Indian described his own experiences as he recalled them during interviews months or years afterward. Placed together and allowing for individual view positions, the Indian descriptions form a mosaic quite compatible with the archaeologists' reconstructions.

American First Nations maintaining histories over centuries by trained oral transmission goes counter to standard American history glorifying Columbus and the onset of European conquests. Out in lands officially described as wilderness inhabited only by savages little more than brute animals, lands that English Puritans thought God had prepared for them to civilize, American Indians were called an inferior race well into the twentieth century. Dogma that served conquerors and colonizers should have died with Custer at the Little Bighorn, but battle wages on between respect for Indian people and campaigns to deny their full humanity. Increasingly, archaeology is demonstrating the historical value of First Nations traditions and helping these nations recapture their history on their own terms, rather than from the viewpoint of their conquerors.

Figure 4.5 Battle of the Little Bighorn, drawn by White Swan, a Crow Indian participant. In contrast to Euroamerican painters glorifying "Custer's Last Stand," White Swan depicts himself on the U.S. side in the battle against his Sioux and Cheyenne enemies. *Courtesy of the Autry National Center, Southwest Museum, Los Angeles. Photo #609.G.185.*

SOURCES

Bernardini, Wesley, 2005. *Hopi Oral Tradition and the Archaeology of Identity*. Tucson: University of Arizona Press.

Brownlee, Kevin, and E. Leigh Syms, 1999. *Kayasochi Kikawenow: Our Mother from Long Ago (An Early Cree Woman and Her Personal Belongings from Nagami Bay, Southern Indian Lake)*. Winnipeg: Manitoba Museum of Man and Nature.

de Laguna, Frederica, 1960. *The Story of a Tlingit Community: A Problem in the Relationship between Archeological, Ethnological, and Historical Methods*. Smithsonian Institution, Bureau of American Ethnology Bulletin 172. Washington, DC: Government Printing Office.

—. 2000. *Travels among the Dena: Exploring Alaska's Yukon Valley*. Seattle: University of Washington Press.

Ferguson, T. J., and Chip Colwell-Chanthaphonh, 2006. *History Is in the Land: Multivocal Tribal Traditions in Arizona's San Pedro Valley*. Tucson: University of Arizona Press.

Fox, Richard Allan, Jr., 1993. *Archaeology, History, and Custer's Last Battle: The Little Big Horn Reexamined*. Norman: University of Oklahoma Press.

Kerber, Jordan E., editor, 2006. *Cross-Cultural Collaboration: Native Peoples and Archaeology in the Northeastern United States*. Lincoln: University of Nebraska Press.

Lyons, Patrick D., 2003. *Ancestral Hopi Migrations*. Tucson: University of Arizona Press.

Parezo, Nancy J., editor, 1994. *Hidden Scholars: Women Anthropologists and the Native American Southwest*. Albuquerque: University of New Mexico Press.

Ridington, Robin, and Dennis Hastings (In'aska), 1997. *Blessing for a Long Time: The Sacred Pole of the Omaha Tribe* . Lincoln: University of Nebraska Press.

Scott, Douglas D., Richard A. Fox, Jr., Melissa A. Connor, and Dick Harmon, 1989. *Archaeological Perspectives on the Battle of the Little Bighorn*. Norman: University of Oklahoma Press.

Thomas, David Hurst, 2000. *Skull Wars: Kennewick Man, Archaeology, and the Battle for Native American Identity*. New York: Basic Books.

Wall, Diana diZerega, and Anne-Marie Cantwell, 2004. *Touring Gotham's Archaeological Past*. New Haven: Yale University Press.

Watkins, Joe, 2000. *Indigenous Archaeology: American Indian Values and Scientific Practice*. Walnut Creek, CA: AltaMira.

Zedeño, María Nieves, 2007. Blackfeet Landscape Knowledge and the Badger-Two Medicine Traditional Cultural District. *SAA Archaeological Record* 7 (2):9–12, 22.

5 Finding Diversity

"Man" used to be what archaeologists looked for, a faceless *Homo sapien*. Then, with the U.S. Civil Rights Act of 1964 making it illegal to discriminate against people on the basis of their race, ethnicity, religion, or gender, and former colonies of European empires gaining independence, the idea of respecting diversity among humans gradually gained force. Instead of forcing all children to practice Christianity and learn how superior European civilization had been to all others, Americans began to understand cultural relativism, the principle basic to anthropology that each society evolved a culture meeting its people's needs. Archaeologists realized that cultural diversity was a feature of the past that could be related to the present. They also became aware that their narrow focus on artifacts—material things—hid much of the social life in the past—for example, the roles expected of women, cultural differences between different ethnic groups, the presence of social classes, and the lives of ordinary working people.

What bothers people when archaeologists emphasize a feminist or class-conscious or ethnic standpoint is the challenge to the comfortable modern notion that humans are basically all the same, except some were or still are "primitive" and we are civilized. When anthropologist Anthony F. C. Wallace insisted, at the beginning of the 1960s, that societies are not homogeneous culturally but instead, organized diversity, he reminded us that, at a minimum, there will be the diversity of men and women, old and young and infants, healthy and weak. Add to that the social teaching that some people are "naturally" superior, born to "better" families (even in small hunter-gatherer communities, where such families train their children to be leaders). Archaeologists wanting to be known as scientists tried to identify patterns of organization in the several types of economies they classified, that is, to discover laws of or regularities in the organization of basic diversity.

Up until the 1960s, the goal of archaeology was to dig deep five-foot-square pits to recover series of potsherds and stone "projectile points" constituting a region's "culture history" (sequence of societies, and their cultures, through time). Then came the touted revolutionary New Archaeology, the processual archaeology we described in Chapter 2. Young men following charismatic Lewis Binford looked like 1960s hippies with long hair, beards, and shapeless khakis, but their credo was more fascist[1] than socially radical: they wanted to reveal universal scientific laws governing human behavior. Binford promised "that data relevant to most, if not all, the components of past sociocultural system [*sic*] *are* preserved in the archeological record" (Binford 1978:22; his emphasis and spelling). "Sociocultural systems" were diagrammed as boxes linked by flowchart arrows, archaeological artifacts and features were put into the boxes, and ecological constraints were invoked as limits. Statistics became popular, because the advent of computers made it easier to use statistics. The crux of the New Archaeology was its "hypothetico-deductive" procedure, wherein the archaeologist states a hypothesis and then seeks data to either confirm or negate the hypothesis. Originating with philosophers concerned with the physical sciences, hypothetico-deductive logic was rejected by historians and most mid-century archaeologists because human behavior is more complex than the behavior of molecules and subatomic particles. Even the behavior of atoms is described by today's scientists as *observed regularities* rather than "universal laws."

Muffled by the loud preaching of the self-named New Archaeologists was a real shift in who became archaeologists. Until late in the twentieth century, nearly all professional archaeologists were White men. Women who tried to enter the profession were pushed out of fieldwork into laboratory analysis; those who rebelled were seldom given substantial project grants or influential academic posts. Women were denied jobs on field crews—I well remember earning $18 a week as a field assistant when male fellow students lacking my experience were paid $40 on projects that didn't hire women. Professors routinely recommended men students for positions, ignoring women graduates. Thanks to the 1964

[1] The fascist angle is that from the 1920s to the 1950s, many Americans and Europeans were attracted to an anti-Communist movement advocating a strong centralized government emphasizing highly regulated control, supposedly derived from scientific knowledge. Social scientists would be recruited to formulate laws of human behavior and advise governments to operate in conformance with these fundamental directives. The "counterculture" radicals of the 1960s were reacting to 1950s conservatives promoting the discovery of and regulation by such postulated laws.

Civil Rights Act and the feminist movement of the 1970s, more and more women entered the profession, they documented and protested inequities in funding and hiring, and, their consciousness raised, they pondered upon women's roles, oppression, conflict, and bias.

During this time, ethnohistory developed into a recognized subfield of history calling upon anthropology and archaeology. Ethnohistorians viewed documents from the standpoint of their subjects, such as American Indians, slaves, or women, rather than that of literate White men writers. The impetus for ethnohistory as counterpoint to standard academic history came from the U.S. Indian Claims Commission, created soon after World War II to compensate First Nations for territory seized without just payment. Structured by a lawyer, the Commission operated like a court of law, listening to pro and con arguments from lawyers representing the two sides, either the U.S. government or Indian tribes. Both sides hired researchers to establish First Nations' territories, their treaties with the United States, and their subsequent histories. Some cases called up archaeologists to testify on sites linked to plaintiff tribes. The Commission was about injustice, and it heard a great deal regarding injustice. Simultaneously with the hearings, Indian men and women who had returned from serving in the armed forces pushed for better opportunities in civilian life, more respect. Radicals organized the American Indian Movement (AIM) in 1968, staging dramatic protests.

Out of women's raised consciousness, ethnic groups' protests against discrimination, and disillusion over America's prosecution of war in distant Vietnam, North America shifted away from tacit acceptance of Anglo-American dominance. Canada began to picture itself as a cultural mosaic. Archaeology, although it ostensibly deals with the past, reflects contemporary attitudes, so as society openly recognized value in segments of our society previously held in low esteem, archaeologists looked for evidence of these groups. Gender archaeology, the archaeology of slavery, Marxist archaeology exploring class conflicts, and more respectful engagement with First Nations grew out of a heightened realization that, as anthropologist Anthony Wallace put it, all human societies represent "organization of diversity," not mechanical "replication of uniformity."

■ ■ ■ GENDER ARCHAEOLOGY

New Archaeologists had a box in their ubiquitous flow charts for "sexual division of labor." They inferred—oh, excuse me, they hypothesized and deduced—the presence and activities of women from artifacts ethnographically associated with women's tasks—for example, corn-grinding and hide-tanning.

Of course, they may have uncovered graves with female skeletons, supporting an association of certain artifacts with women, although a disconcerting number of prehistoric graves have a skeleton of one sex buried with artifacts conventionally assumed to relate to the opposite gender.[2] A "sexual" (gender) division of labor was taken for granted, and further, it was presumed that men went out hunting, and women stayed close to home doing routine household maintenance and childcare. Coincident with the formulation of the New Archaeology, a major conference was held on the topic "Man the Hunter," to work out "a general statement about the ecology, subsistence, and social organization of contemporary hunting and gathering peoples" (Lee 1968:344). Considering that hunting and gathering peoples live on the equator and near the North Pole, in tropical rainforests and deserts and northern coniferous forests and tundra, it is extraordinary that anyone thought they could lump them all under a general statement about *the* ecology, *the* subsistence, and *the* social organization of hunter-gatherers. That hope illustrates the premise commonly accepted by the New Archaeology movement in the 1960s, that human behavior could be reduced to a few basic forms linked to stages of cultural evolution, subject to universal laws of nature. For them, discussion of gender roles was "noise" obscuring underlying laws (or from more sophisticated academics, "regularities") such as men hunting and dominating societies, women dependent and subordinate.

The times they were a-changing. Women read journalist Betty Friedan's *The Feminist Mystique* (1963) and realized they were born as human beings, not housewives. Intellectual women read French philosopher Simone de Beauvoir's *The Second Sex* (1953) and learned there is a long history of denying women opportunity to use all their talents. Some women went further, insisting that women once ruled societies, symbolizing women's power in the image of a great Goddess, and accounting for historical circumstances by supposing war-loving men conquered women's peaceful societies and elevated, over the Goddess, their wrathful masculine God (see Chapter 6). Even women who didn't read much or question the dominance of men in churches and business paid attention to feminists pointing out unfair pay scales, including in archaeology, where women automatically were paid less than men in similar jobs.

In the 1980s, as a steadily increasing number of women obtained professional positions in archaeology, they began to notice a male bias not only in the

---

2 Skeletons are biologically male or female sex; artifacts may be associated with *gender* roles. Transvestites are biologically one sex but pass as persons of the other gender; that is, gender is a societal category but sex is a genetic characteristic.

tendency to hire and to fund men, but also in the interpretation of archaeological data. Although the "Man the Hunter" conference failed to synthesize ethnographies into a general statement about *the* subsistence and *the* social organization of hunter-gathering communities, it had rocked anthropology by bringing out observations demonstrating that for most hunter-gatherers, women's gathering of plant foods and snaring of smaller game and fish produced the bulk of food consumed and the daily staples (Figure 5.1). In other words, Paleolithic hominins and Paleoindians very likely depended on their womenfolk for survival. Big game meat was a treat. Forget the husky guy in a loincloth brandishing a spear at a mammoth; the common scene was a weary fellow grateful that his missus had fish on a spit, or a rabbit, or wild potatoes stewing with herbs. Archaeologists should pay more attention to hearths and kitchen tools. Raised consciousness over women's presence in sites stimulated the invention of techniques for recovering a full range of plant and small animal foods, mainly through floating excavated soil in tanks so that miniscule remains—bone fragments, seeds, bits of wood or fruit rind—could be skimmed or filtered out. A few researchers looked at bone tools for evidence for basketry and weaving. It all came to a head in 1989 when the University of Calgary's Archaeology Department sponsored a conference on "The Archaeology of Gender." Several hundred scholars and students came, contributing papers to a fat volume.

From that point on, dozens of conferences and meeting symposia focused on identifying, rather than simply assuming, a sexual division of labor, and concluding that it is better conceptualized as *gender* roles, since biological males may perform "women's" tasks—the men may be slaves set to household drudgery, or might feel a vocation, perhaps spiritually directed, for such tasks. Similarly, there are plenty of historical and ethnographic cases of women warriors, hunters, traders, high-ranking priests, and rulers (think Hatshepsut, an Egyptian pharaoh, or Queen Elizabeth I). In some societies, men are supposed to be the ones who knit, weave, or make pottery. Women may chip stone tools: hide-scraper blades, knives, and hoes are quickly worn down, and women cannot wait helplessly until a man comes to replace their tools. As obvious as this seems, in the 1980s a few of our macho flintknapping male archaeologists insisted that only men can chip stone tools. And all the stone blades around hearths associated with houses and bone butchered for cooking, they're all "projectile points," not kitchen tools? Those macho guy archaeologists didn't notice that the pointed stone blades in the domestic debris are shaped like the blades of our small kitchen knives (these guys didn't spend much time in their wives' kitchens) and that shape would be ill-balanced for projectile points.

FIGURE 5.1 Paiute women gathering food, Utah, 1870s. Ethnographic descriptions of hunter-gatherer societies make it clear that "Man the Hunter" depended daily on the knowledge and work of "Woman the Gatherer" (and that men, too, gather plant foods and medicines). *Courtesy Smithsonian Institution National Anthropological Archives.*

Gender archaeology drew attention to data that formerly might not have been commented upon. For example, in a three-thousand-year-old cemetery in Wisconsin, its interments saved by archaeologists before imminent destruction, expensive copper artifacts and a block of obsidian glass imported over a thousand miles from Wyoming were placed in graves of young women and children, more than in men's graves. Two thousand years earlier, graves in the same region had copper goods and other artifacts interred with most of the dead. Probably the more ancient people shared more equably within the community, while the later society singled out young women and children for special status. But what status? Controversies erupt over data like these. The archaeologist who wrote his dissertation on the material inferred that the community may have prized women reproducing their people, and the children who had been their hopes for the future. Some of the young women likely died in childbirth, a frequent danger everywhere before the twentieth century. Other inferences are possible, too. Thorstein Veblen, an American economist from a century ago, wrote of how nineteenth-century middle- and upper-class women were "put on a pedestal," loudly admired but ignored when important decisions were made. Possibly, but *not* probably in this Midwestern Early Woodland society of 1000 B.C.E., young women and children were sacrificed in rituals and buried with gifts for the gods. If the graves were in Mexico and many centuries later, it would be reasonable to infer that the maidens and children had been bedecked with expensive ornaments and then ritually killed to supply the needs of deities. It is more probable that the Midwestern women and children, like those Veblen observed, were members of respected families who marked higher social status by bestowing expensive goods on their loved ones.

Another case reveals how desperately a few scholars clutch their accustomed ideas. A burial in a wooden tomb under a mound at Vix in Burgundy, France, dated to about 500 B.C.E. (contemporary with Classical Greece), held the body of a woman in her early thirties, wearing a fancy gold neck ring imported from Spain, amber and jet bead bracelets, and fine bronze ornaments. She lay on an elaborately decorated wagon, its wheels removed and propped against a wall. Across the chamber, occupying almost as much space as the wagon, was an enormous bronze wine bowl five feet (1.64 meters) tall, brought from the Mediterranean, accompanied by a bronze pitcher, a silver and gilt bowl, and lovely Greek-style drinking cups. What would you infer to be the best explanation? Most of us say this was a queen or royal princess whose subjects believed she merited a lavish outlay of luxuries even in death. A very cautious archaeologist, remembering Veblen, could consider the possibility that the woman was a servant dressed to please a god or ancestor, sacrificed with the superhuman wine set. But a couple of older German archaeologists couldn't accept the idea that their ancestors might have honored a woman with the richest array of goods known from the Iron Age. Their interpretation was that the biological anthropologist who had analyzed the skeleton as female had mistaken a slightly built Mediterranean transvestite priest for a woman—never mind that no one has heard of such characters from Germanic traditions or later Roman observations of their northern neighbors. Such an elaborate interpretation—"slightly built" "Mediterranean" "transvestite" "priest"—surely needs Ockham's Razor to slice away the speculations!

■ ■ ■ ARCHAEOLOGY OF SOCIAL CLASS

As the richer concept of cultural variations in gender roles superseded assumptions of a universal sexual division of labor, more dimensions came into discussion. What about social classes? leadership and subordination? A symposium on "Manifesting Power" mentioned that, as economist Thorstein Veblen wrote, the really rich don't flaunt it.[3] Wealth objects don't invariably link to power. Signals of social class can be subtle; people born into the upper class don't wear "bling" jewelry like that favored by gangsta rappers.

Underlying the late-twentieth-century turn toward revealing the circumstances of the lower classes was the shift from museum-sponsored projects expected to bring back beautiful objects for display, to grant-supported investigations of past ways of life and contract archaeology assignments to find out what had occupied sites slated for development. Thinking about social classes spurred efforts to locate and excavate commoners' homes, even brothels in urban historical projects. Brothels conveniently located in Washington between Capitol Hill and the White House? The archaeologist conducting cultural resource management excavations prior to new buildings there across from the Smithsonian Mall consulted nineteenth-century newspapers, censuses, and city directories: she could have politely identified the houses as "females' boarding-houses" but she chose to be honest. Inference to the best explanation would have concluded that, too, in light of the abundance of bones from steaks and beef roasts—virtually absent from trash of respectable working-class households in the neighborhood—and a lot of broken glass from light fixtures, evidence that the inhabitants of the brothels worked at night. In her report, the archaeologist emphasizes that prostitution paid a living wage to a nineteenth-century woman, unlike other working-class occupations hiring women, and that for most it did not preclude marriage and homemaking later.

Another study of working-class men and women drawn from a controversial situation involved families in early-twentieth-century mining camps in Colorado.

---

[3] An English archaeologist friend whose son won a scholarship to the prestigious English prep school Eton apologized for picking him up at the end of term in the family Volvo. "No, Mum," he replied, "that's what the earls and dukes drive. Those Jaguars and Mercedes belong to the 'nooves'—nouveau riche. We laugh at them."

"Country Wives" in the Fur Trade

Ethnohistorians in Canada found that the story of the fur trade, so important in the development of their country as well as the United States, is much more than an adventure of hardy White men paddling canoes deep into uncharted wilderness, bartering trinkets to naive savages for their beaver-skin robes. In the first place, European traders entered into a continent-wide indigenous commerce millennia old, carried along well-known transportation routes. Fine furs and useful hides were already being traded, along with inorganic raw materials, finished products, food, and slaves. To make a profit in spite of the enormous transport costs between Europe, China, and America, European fur traders needed to find goods their American customers wanted, goods cheap to produce in Europe and economical to ship. Metal knives and kettles, sewing awls, colored glass beads, sheetmetal for cutting out arrowpoints and dance tinklers, strike-a-lights, wool cloth, and liquor fit the bill as useful or desired products cheaper for Indians to buy than to make equivalents with their own technology. European traders had the odd notion to build wooden habitations like those in northern Europe rather than live in American-style wigwams or tipis among their customers. Consequently, archaeologists can distinguish remnants of trading posts from indigenous settlements, although trading posts tended to be short-lived, posing challenges to link particular remains to one or another post documented in an area.

According to popular history, White men lived in the trading posts and Indians camped outside. Ethnohistorians carefully analyzing traders' reports and explorers' journals realized that the "Canadians" paddling freight canoes could be French Canadians, British, Iroquois Indians, or sons of any of these and indigenous women along the long river routes. As at Fort Ross in California, people of mixed parentage became a recognized class, in Canada termed Métis (French for "mixed"). European men in the fur trade took local women as common-law wives *à la façon du pays,* "according to the custom of the country"—hence the women were termed "country wives" (Figure 5.2). Some traders brought their country wives and children back with them to Eastern cities when they retired from trading posts; others terminated the marriage, leaving the women and children with their indigenous relatives. (There were also an untold number of children from one-night stands, drunken debauches, and heartless seductions.) Does the archaeology of fur trade posts reveal the diversity recorded in documents?

I excavated a post dating 1768–1773, built on the south bank of the Saskatchewan River near the present town of Nipawin, in central Canada (Figure 5.3). The traders were partners James Finlay, a Scots immigrant businessman from Montreal, and François Le Blanc, an experienced *voyageur*. At the time,

FIGURE 5.2 Cree "country wife" in the Canadian fur trade, in this picture with a Cree man. *Painting by William Richards, © Hudson's Bay Company Archives.*

England's Hudson's Bay Company held a monopoly on fur trading in interior Canada; Finlay and François' enterprise was illegal. If they kept records, they haven't been located. Historical documentation of the post comes from employees of the Hudson's Bay Company spying on the competition. From them, we know that François was an unprepossessing middle-aged man more friendly to his Indian customers than was customary in Hudson's Bay Company posts, and that he had a wife and son with him. A descendant researching the Le Blanc family told me that the wife who was with François then was his second wife, whom he married after he was widowed, and she was the daughter of a family at the major trading depot Michilimackinac—very possibly a Métis girl. Finlay had a Eurocanadian wife back in Montreal and brought an Anishinabe country wife to the post (which was in Cree country). Sons of both his wives worked in the fur trade early in the nineteenth century. We lack descriptions of the men employed at François' House, as the post was called, or of women who may have lived with them.

Indigenous manufactures recovered from François' House included sherds of small clay pots, stone knife and scraper blades, a drill bit, mauls, bone awls, hide-scrapers, a harpoon, beads, a thong softener, an antler flaker for making stone artifacts, stone pipe bowls, and shell beads. Some of these objects lay discarded in trash pits, others clustered against the outside of the post's stockade wall, especially near the sunny southwest corner facing the river. I infer from these clusters that the country wives living in the post sat in this pleasant corner processing food (stone knife blade, sherds), perhaps sewing or making baskets or birchbark containers, chatting as little Jaco Finlay and the other children played near them. The clusters of sherds and stone blades were so close against the stockade wall that it seems improbable they were discarded by Indian customers camped near the post. If the interpretation is correct that the clusters represent Métis and Indian country wives, it is interesting that they preferred to use traditional implements they were accustomed to from their mothers' work, rather than foreign ones imported by their husbands. Experiments have demonstrated that indigenous stone and bone artifacts generally function as well as eighteenth-century traders' metal imports, the principal advantage of the imports being that they lasted longer.

FIGURE 5.3 Alice Kehoe recording fallen chimney, François House fur trade post, Saskatchewan. The first successful trading post on the Canadian prairies, 1768, it was built by a French-Canadian *voyageur* in partnership with a Scots merchant in Montreal. Both partners had their wives in the post. *Photo by Thomas F. Kehoe.*

Factoring diversity into an archaeological scenario not only furnishes alternate and enriching interpretations of data, it also points up taken-for-granted aspects. Why should the men at François' House work so hard to fell trees, drag them to the clearing, lop off branches, dig a wall trench, set up the logs to make cabin and stockade walls, top the upright posts with log wall-plates, lay roof timbers, cut wood shakes, nail them on, chink everything with clay daub? The country women could have quickly erected comfortable tipis. François didn't really need a stockade, except to keep bears out, since he welcomed his customers into his post. Why did he, or Finlay, carefully pack and carry thousands of miles a nice white teacup and several glass bottles of Turlington's Balsom, an English patent medicine? While indigenous traders were hosted in their customers' homes, Europeans signaled their difference by building rectangular wooden structures and by their dependence on their own manufactures, portaged on their backs around rapids, shoals, and land connections between waterways. From the standpoint of efficiency and energy costs, European traders were crazy—did a country wife occasionally wonder about her man and his comrades?

Their Métis children grew up creating a Métis culture with components from both parent cultures, able to live off the country and dance to French fiddles, wintering in cabins, traveling to the hunt in high-wheeled carts, speaking a creole language combining French and Cree. It was a successful adaptation for two generations, until an expanding Canadian government imposed its rigid land-parceling property laws, settlers moved west to farm, and the bison herds supporting Métis as well as Indians were exterminated. From burned log posts, thousands of tiny beads lost in floor sweepings, sherds, rusted knife blades, patent medicine bottles, butchered elk, moose, beaver, and waterfowl bones, and clay pipes—the inventory of an early prairie-border trading post—the archaeologist sees the genesis of the Métis Nation in ethnic and gender diversity at the site.

In 1913, coal miners and their families on strike against Colorado Fuel and Iron Company camped in tents after being evicted from company houses. Strikers protested unsafe conditions in the mines, low wages, high rents charged by the company for its houses, and high prices in its stores where miners were forced to shop. The strike went on for months, until in 1914 the Company broke it by assaulting the Ludlow tent camp, killing seven men, two women, and eleven children. A Marxist-inspired archaeological project compared remains of miners' houses from before and after the Ludlow Massacre, as the strikebreaking assault has been called. Before the strike, miner families eked out a bare living by taking in single men as boarders, cramming as many as twenty people into a four-room house. Miners' wives cooked for everyone, and the archaeologist found that many purchased commercial-size cans of food to do so. After the bloody end of the strike, Colorado Fuel and Iron tried to improve its public-relations image by building dormitories for single workers and discouraging families from boarding the men. Wages were raised, but inflation negated the purchasing power. Excavations of post-Massacre house sites showed that miners' wives resorted to home canning, evidenced by quantities of broken Mason jars and lids, to feed their families; they no longer needed, nor could they afford, to buy canned food.

Analyzing data from several bars in a frontier town revised stereotypes about pioneer settlers; contrary to the assumption that the frontier had no social classes, the bars clearly drew different classes of patrons. The Nevada State Historic Preservation Office was concerned, in the 1990s, that tourist development in picturesque Virginia City was obscuring its nineteenth-century history. Archaeologists chose documented locations of four saloons to excavate: a bar connected to Piper's Opera House, a saloon catering to African-Americans, and two operated by Irishmen in a disreputable part of town (Figure 5.4). As expected, the Opera House bar was embellished with gold-leaf-accented wallpaper and served patrons with crystal stemware; the two Irish

FIGURE 5.4 Excavation of a saloon in Block 88, Virginia City, Nevada. Archaeologists excavating sites of several saloons in the frontier town revealed a diversity of social class and race, contrasting with the stereotype of Wild West saloons in films and television. *Photo by Julie Schablitsky.*

saloons were much simpler, although one had a few crystal wineglasses; and the African-American's saloon was in between in quality of furnishings. Given the discrimination suffered by Black people even on the frontier, it wasn't surprising that this saloon had dinnerware to serve food as well as liquor to its customers, who probably weren't allowed to dine in White restaurants. Contrary to Hollywood movies, shoot-outs apparently weren't common in Virginia City's saloons, judging from the small number of cartridge shells found by the archaeologists in three of the sites. The fourth had more than a thousand shell casings, but not from gunfights—it housed a Shooting Gallery where customers could safely target-shoot. This establishment's site also yielded relics of women's fancy clothing, hinting at prostitutes, while DNA analysis of cells on a clay tobacco pipe stem revealed an African-American woman had enjoyed a pipeful there.

ARCHAEOLOGY OF ETHNIC GROUPS

Before the Civil War, the majority of African-Americans were not free to relax in saloons. Excavations at both George Washington's home during his presidency, and at Thomas Jefferson's Monticello, Virginia, estate revealed small cabins where the slaves of these presidents lived, and at each fine house, a tunnel connected the slaves' quarters to the master's house. Why a tunnel? So that the wealthy family and their guests would not see the slaves coming and going to serve them. We cannot pretend that our first and third presidents didn't see the contradiction between the revolution they led for liberty and their exploitation of Black people, when they built tunnels so they could avoid seeing their enslaved workers. In 2007, the National Park Service faced a difficult decision: should it incorporate the slave tunnel with the restored house, adjacent to the Liberty Bell and Independence Hall in Philadelphia, where Washington lived in the 1790s? Or should they cover over the excavation and present a bland exhibit on the genteel life enjoyed by the Washingtons? A Philadelphia civic leader urging honesty told journalists viewing the excavation, "As you enter the heaven of liberty, you literally have to cross the hell of slavery."

Near the World Trade Center in New York, builders came upon a cemetery devoted to eighteenth-century African-Americans, both free and slave. Most of the bodies were removed for the office-building construction, but under protest from New York's African-American citizens, a portion of the cemetery was preserved. Archaeologists and biological archaeologists analyzing the removed skeletons and their associated artifacts learned much about the diverse origins of these Black New Yorkers, and how they amalgamated bits of African cultures with their eighteenth-century colonial American life. A couple of older adults had teeth filed to be decorative, indicating they must have grown up in Africa. One of these, a man, showed in his bones that his childhood had been comfortable, while in his adult life he had performed such hard labor that his vertebrae had cracked—literally back-breaking labor. On his wooden coffin was a design similar to one commonly used by the Akan people in Ghana, suggesting he had been captured there and brought to America as a young man. A sad commentary on African-Americans in eighteenth-century New York was that the cemetery had a high proportion of infants and young children, high even for that time (before modern medical care), and they and the young adults showed in their bones the stress lines indicating poor nutrition and/or illnesses, while many of the older adults, like the man with filed teeth, did not show childhood stress. It was general among the adult skeletons to see severe muscle stress on bones, indicating hard labor. The African-American Burial Ground in lower Manhattan is mute testimony to the cruelty of slavery.

California has been a land of ethnic diversity for millennia. At European colonization in the late eighteenth century, indigenous peoples lived in independent communities managing resources through regular controlled burnings of the land, seeding and limited cultivation of food plants, fishing and shellfish collection, and hunting both land and sea mammals and birds. Landscapes of meadows and open woodlands depended on Indians' scheduled burning and so were disappearing in the twentieth century, replaced by scrub, due to the Euroamerican policy of extinguishing wildfires. Because Californians' time-honed methods of sustaining resources were so different from European agriculture, not until recently and through years of debate has the extent and value of indigenous management been acknowledged. Although California had a greater number of indigenous languages than any other region, its peoples traded widely, using shell units of currency, and knowledge of resource management practices was widespread. Colonization—first by Spanish Mexican Catholic missionaries in the late eighteenth and early nineteenth centuries, then by Mexican ranchers taking over mission properties and enslaved Indian laborers, and finally, after 1846, by Americans—severely reduced the region's First Nations. Some fled into the Sierras, constantly on the run from trigger-happy gold miners and cowboys; others became peons on farms and ranches. Indian men and women married fellow laborers—Mexican, Filipino, Hawaiian, Asian, African- and Euroamerican. Quite a few Indian peoples nevertheless persevered, turning a big corner economically—and as a result, politically—when in the late twentieth century several followed the lead of Seminole and Oneida First Nations in opening high-stakes casinos attracting tourists.

California archaeologists once tended to be rather dismissive of their state's indigenous past, empty of ruins or pots (finely woven baskets, which don't preserve into the archaeological record, were used where other cultures used ceramics). The post-World War II Indian Claims Act, designed to settle First Nations' grievances against the federal government, engaged anthropologists to investigate and document Indian practices and territories, and these researches illuminated their clients' sophisticated resource management. The older generation of anthropologists—who saw only "tribelets" with "cults," hunting, fishing, and gathering nature's bounty—passed away, leaving the field (pun intended) to those who could more sincerely appreciate alternative lifeways. It is surely not coincidental that this appreciation developed along with more nuanced conservation movements, New Age religion, and beaches crowded with bikini-clad sun-lovers wearing less than the "naked savages" attacked by nineteenth-century colonists. By late in the twentieth century, Californians had discovered the wisdom of their predecessors.

Historic archaeology in California had focused on restoring the Spanish missions for tourist visits. With the state's population increasingly so visibly diverse, archaeologists began in the 1980s to investigate the state's colonial Mexican ranchos and sites settled by Russians or Chinese. One of the largest projects was at Fort Ross on the northern California coast. Despite its English name, Fort Ross (a corruption of Fort Russia) was an outpost of Russia's fur trade from 1812 to 1841. (Recall that Alaska at that time was called Russian America. Years later, in 1867, Russia sold Alaska to the United States.) Fort Ross was a substantial settlement run by the Russian-American Company traders to collect sea-otter skins (Figure 5.5). Alutiiq Indian men skilled in hunting the fast, intelligent animals from kayak-like baidarkas were forcibly brought in from Alaska (that is, Russian America), and Siberian native men, Hawaiians, and local Kashaya Pomo and Miwok Indians were also employed in this ethnically diverse trading post. Many of the foreign men married Pomo women, producing a bilingual generation that remained after the Company shut down the post. Archaeological excavations exposed a mix of material cultures reflecting the ethnic mix, and showed that few of the imported Russian manufactures were given or sold directly to the native people. Instead, the imports were sold to Spanish-Mexican ranchers and missions, with the Indian employees and wives around the trading post getting imported items as second-hand used goods. Rather than pay sufficient wages, the Russian American Company expected Indian wives of their laborers to support themselves through traditional food-getting and crafts, resulting in many of the second-hand imports being reworked into Indian-style artifacts and the families having little choice in regard to maintaining Indian ways. Fort Ross is now a State

FIGURE 5.5 Fort Ross, the Russian-American Fur Company post in northern California. Fort Ross was the southernmost of the Russian-American Company's several posts along the northern Pacific coast. It housed Russians, Siberians, Aleuts, and Hawaiians, with many marrying local Pomo and Miwok Indian women. *Photo M. Allen.*

Historic Park displaying the diversity to thousands of visitors annually. Today's Kashaya Pomo live close by, as they did during the trading post's heyday, and participate in Park activities. With limited written records from Fort Ross, archaeology has helped to untangle the complex web of cultures surrounding the fort and reclaim the history of the indigenous inhabitants.

While Fort Ross was operating in northern California, tying Russia, Siberia, Alaska, Hawaii, and California's Indian and Spanish-Mexicans into a remarkable ethnically diverse web of hunters, traders, laborers, and their wives and families, Canada and much of the United States of that time had its own fur trade, based on beaver rather than sea otter. One of the sites I've excavated was a Canadian fur trade post, not as ambitious or diverse as Fort Ross, but like it, showing that native wives of foreign traders tended to maintain their own cultural traditions even within their husbands' cabins.

■ ■ ■ THE DIVERSE STUDY OF DIVERSITY

The cases discussed in this chapter are examples of diversity within a site. It is not uncommon for sites with a long history to have series of occupations, representing a diversity of cultures through time. For example, for several thousand years, eastern Mediterranean ports like Tyre or Byblos, now in Lebanon, mixed groups of merchants, sailors, and people in service occupations from a number of countries, and the archaeology of these ports and islands is further complicated by their being conquered by a succession of empires, from Egypt, Phoenicia, Persia, Macedonia, and Rome to Venice and then Ottoman Turks, Britain, and France. When a government wants to attract tourism to one of these places, it may provoke citizen protests by selecting only one portion of the millennia of occupations. Should Crete, for instance, showcase its Bronze Age Minoans of three thousand years ago, or its fifteenth-century Venetian architecture, or its quaint, modern Greek fishing villages? Restoring one era's ruins likely destroys earlier or later structures on the site. This is a serious dilemma, and controversial when archaeological excavations focus on a single time period.

Some archaeologists continue to search for patterns that indicate regularities in human societies' responses to environmental and demographic situations. Others try to identify "agency" from archaeological data, meaning the effect of individuals'—*diverse* individuals'—choices and actions. Instead of thinking of faceless *Homo sapiens* existing in one or another cultural "stage," these archaeologists try to give us glimpses of actors in busy scenes. Those people in the past weren't robots hunting or gathering or hoeing crops, those people made decisions, tried to manipulate others; there were moral dimensions to their lives, and an archaeology of diversity can sensitize us to that, and to the moral implications of our own work. In the next chapter, we'll read about encounters of archaeology with religious ideas, when questions of implications may seem, to some, to threaten their faith.

Sources

Anderson, M. Kat, 2005. *Tending the Wild: Native American Knowledge and the Management of California's Natural Resources*. Berkeley: University of California Press.

Anthony, David W., 2007. *The Horse, the Wheel, and Language: How Bronze-Age Riders from the Eurasian Steppes Shaped the Modern World.* Princeton: Princeton University Press.

Arnold, Bettina, 1991. The Deposed Princess of Vix: The Need for an Engendered European Prehistory. In *The Archaeology of Gender: Proceedings of the 22nd Annual Chacmool Conference*, edited by Dale Walde and Noreen D. Willows, pp. 366–374. Calgary: Archaeological Association of the University of Calgary.

Binford, Lewis R., 1968. Archeological Perspectives. In *New Perspectives in Archeology*, edited by Sally R. Binford and Lewis R. Binford, pp. 5–32. Chicago: Aldine.

De Beauvoir, Simone, 1953. *The Second Sex*. London: Jonathan Cape.

Dixon, Kelly J., 2005. *Boomtown Saloons*. Reno: University of Nevada Press.

Friedan, Betty, 1963. *The Feminine Mystique*. New York: W. W. Norton.

Herzfeld, Michael, 1992. Metapatterns: Archaeology and the Uses of Evidential Scarcity. In *Representations in Archaeology*, edited by Jean-Claude Gardin and Christopher S. Peebles, pp. 66–86. Bloomington: Indiana University Press.

Kehoe, Alice Beck, 1991a. The Weaver's Wraith. In *The Archaeology of Gender,* edited by Dale Walde and Noreen Willows, pp. 430-435. Calgary: Archaeological Association, University of Calgary.

—. 2000a. François' House, a Significant Pedlars' Post on the Saskatchewan. In *Material Contributions to Ethnohistory: Interpretations of Native North American Life* , edited by Michael S. Nassaney and Eric S. Johnson, pp. 73–187. Gainesville: University Press of Florida and Society for Historical Archaeology.

Kehoe, Alice Beck, and Thomas C. Pleger, 2007. *Archaeology: A Concise Introduction.* Long Grove IL: Waveland Press. Chapter 4 describes Kehoe's career, illustrating difficulties encountered by women archaeologists.

Lee, Richard B., 1968. Comments. In *New Perspectives in Archeology*, edited by Sally R. Binford and Lewis R. Binford, pp. 343–346. Chicago: Aldine.

Lightfoot, Kent G., 2005. *Indians, Missionaries, and Merchants: The Legacy of Colonial Encounters on the California Frontiers.* Berkeley: University of California Press.

Seifert, Donna J., 1991. Within Site of the White House: The Archaeology of Working Women. *Historical Archaeology* 25(4):82–108.

Silliman, Stephen W., 2004. *Lost Laborers in Colonial California: Native Americans and the Archaeology of Rancho Petaluma.* Tucson: University of Arizona Press.

Wallace, Anthony F. C., 1962. *Culture and Personality.* New York: Random House. (Reprinted, p. 213, in Wallace, *Revitalizations and Mazeways*, edited by Robert S. Grumet, 2003. Lincoln: University of Nebraska Press.)

Wood, Margaret C., 2002. Women's Work and Class Conflict in a Working-Class Coal-Mining Community. In *The Dynamics of Power*, edited by Maria O'Donovan, pp. 66–87. Occasional Paper No. 30. Carbondale: Center for Archaeological Investigations, Southern Illinois University.

6 Religion and Archaeology

Controversies have raged for three centuries over whether archaeology can contribute to our understanding of religion. Back in the seventeenth century, English gentlemen speculated whether Stonehenge in England was a temple of the Druid priests described by Romans invading Gaul (France) and Britain.[1] Romans had said Druids worshipped in groves of trees, and Stonehenge is a circle of big stones . . . maybe it *represented* a grove of trees? Stonehenge was built, we now know from radiocarbon dates, nearly three thousand years before Romans saw Druids (see Chapter 8). Archaeology has a habit of uncovering discomfiting data. Amid crowds gathered to celebrate summer solstice Midsummer's Eve at Stonehenge, you can't out-argue the true-blue Britons, dressed in long robes, claiming to be heirs of the Druids and rightful priests for Stonehenge. As the man said, "Don't confuse my mind with facts."

This chapter deals with some of the stickiest controversies involving archaeology, because they touch upon people's faith. Myths are created to support a community, a nationality, an ethnic identity, purportedly explaining why the world is as we find it or how a people came to be. Science and history also work to explain how our world came to be, but through verifiable data, while myths are accepted without efforts to test their validity. This may not be a straightforward difference, because modern religious adherents may claim they have solid data proving their belief, insisting that science proves the historical validity behind

[1] Stonehenge may be the most often viewed archaeological site in the world: it comes bundled as a desktop background image with Microsoft Word.

their myths, which nonbelievers are likely to interpret to the contrary. Because religions originated in the past, and are rarely described except in religious documents, archaeology becomes embroiled in these controversies as the one way to test the truth of religious myths. Here, I'll present several controversial claims involving archaeological data—the Goddess movement, shamanism, the Bible, Mormonism—and how archaeologists deal with these claims.

GODDESSES

Some people seek a suprahuman being with qualities matching the worshipper's ideal. Mainstream religions' conventional images of their deity may dismay a seeker, such as a woman who feels alienated by picturing a wrathful masculine god. Devotion to the Virgin Mary could fulfill a Catholic's longing for a sympathetic divine figure. For others, the concept of a pure and mighty Goddess without masculine kin satisfies a deep-felt yearning. With the growth of the feminist movement in the 1970s, many found it alluring to believe in the existence of an early historical period of woman-dominated societies serving a peaceful, equitable, nurturing Goddess.

Marija Gimbutas was an archaeologist beloved by Goddess-worshippers (Figure 6.1). Gimbutas immigrated to the United States from Lithuania as a young woman and became a research associate in Harvard's Department of Anthropology, where she was given a laboratory for her analyses of material she excavated in southeastern Europe. That being the 1950s, men faculty tolerated the dedicated woman but were hardly collegial. Then Gimbutas obtained a position at the University of California, Los Angeles, and drawing on her years of field research, prepared an exhibit and accompanying book on Neolithic figurines and painted pottery, *Gods and Goddesses of Old Europe*. The title was a bit tongue-in-cheek, since the motifs on pottery might have signified cosmological principles such as life-giving rain, and figurines could be household spirits. Gimbutas used her own serious archaeological work and that of others to theorize an Old Europe of six thousand years ago living under the rule of the Goddess, interpreting these figurines and pottery motifs as a symbolic language communicating rituals and principles from this feminist worldview.

A floodgate was opened. A second edition of the book transposed the title into *Goddesses and Gods of Old Europe*. Californian Goddess-worshippers flocked to Gimbutas's lectures, stimulating her to cater to their pleas with subsequent books, *The Language of the Goddess* and then *The Civilization of the Goddess*. At an annual meeting of the American Anthropological Association held in the state, the professor was trailed by a bevy of maidens in flowing robes

FIGURE 6.1 Archaeologist Marija Gimbutas. Her interpretations of Neolithic archaeological data in southeast Europe emphasized recognition of female deities as well as male. *Photo Ernestine Elster.*

and flower garlands, prayerfully clasping their hands together behind their heroine. What could Gimbutas do? The meeting hotel was a public place. To be adored by flower maidens . . . no wonder archaeologists shy away from Goddess-worshippers. While Gimbutas was recognized as a capable archaeologist, few of her colleagues agreed with her interpretation of the data, no matter how large a public following her ideas attracted.

At a "Gender and Archaeology" conference in California, celebrants of the Eternal Female—dressed in miniskirts, high leather boots, low-cut blouses, and lots of heavy jewelry and hair—hissed Professor Cynthia Eller for her calm, sensible, and sympathetic book, *The Myth of Matriarchal Prehistory*, subtitled *Why an Invented Past Won't Give Women a Future*. Eller described how she herself felt attracted to the idea that once upon a time, good women ruled peaceful societies. If only those bloodthirsty horse-riding nomads hadn't swept out of the steppes, early versions of Attila the Hun, banging battle-axes on helpless farmers, we wouldn't have had all these wars and torture and witch-burnings! But wait, Eller cautioned: it's not simply a battle of the sexes. Women can be cruel, as high-school girls know so well, and men have made some admirable contributions to societies—for example, the Constitution and Bill of Rights of the United States. Searching archaeological and Classical documentary sources construed to show Goddess worship and peaceful matriarchies, Eller found the evidence weak; the alleged prehistory is a myth, and a modern myth, at that. Building a movement on a myth draws eager followers, but, Eller warned, only contempt from more hard-headed citizens. Women, and the men who love them, need to know the historical facts about social structures oppressing women in order to rectify injustice. A wished-for past gets no respect.

A famous site in Turkey has long been a focus of Goddess-love and debate. Çatalhöyük (pronounced Sha-tal-hoo-yook) was excavated in the late 1950s and early 1960s. It was an important find because it evidenced a large, urban-looking village, with rows and rows of houses, early in the development of agriculture (that is, the Neolithic period), 7400–6000 B.C.E. At the time of the first excavations, New Age religions and Goddess-worship were becoming more widely known and popular. The excavator liked publicity and made much of female figurines and of bulls' heads embedded in the walls of houses. Particularly alluring was a clay figure, found in a grain bin, depicting a woman seated on a chair (throne?) with her hands resting on a pair of large felines, possibly leopards, standing beside her chair. Both Marija Gimbutas and the excavator, James Mellaart, interpreted the bulls' heads, figurines, and wall paintings inside the houses to indicate a society very aware of masculinity (bulls) and of female power; Çatalhöyük became a pilgrimage site for Goddess followers. A generation later, excavations were resumed at Çatalhöyük by another excavator, Ian Hodder, who paid more attention to numerous depictions of leopards in house decorations. To the dismay of Goddess tourists, he interpreted the bulls' heads as symbols of feasts of meat, and the famous lady with the big cats to possibly have a war trophy or ancestor's skull between her legs, not a newborn baby, and leopards to indicate balanced, equitable gender roles. He noted that most of the clay figurines do not show sex or gender and were discarded with trash, while some distinctly female figurines lay in houses near hearths. Whatever one thinks of leopards as symbolizing "ambiguity," as this "postmodern" archaeologist suggests, the two series of excavations at Çatalhöyük show neither "matriarchy" nor "patriarchy." The archaeology shows an interesting farming town whose people decorated their clean-swept homes with handsome paintings and heads of wild bulls sporting sweeping horns that their menfolk—probably, menfolk—hunted. Writing would not be invented for another three thousand years, leaving us clueless about what Çatalhöyük's art really meant to its people.

At the extreme, Goddess-worshippers see every round hollow thing, such as a round temple, as a symbol of the Goddess's womb, every triangle in rock art or painted on pottery as a pubis, every pair of short parallel lines a vulva. With the objects' preliterate makers long dead, we can't *prove* the Goddess-worshippers are wrong in their interpretations, but conversely, they can't offer more than romantic presuppositions. Unless an archaeologist can show actual links between ethnographic or historically documented symbols and abstract designs as common as triangles and parallel lines, we should not impose interpretation.

Figurines of women found in Paleolithic sites are popularly assumed to indicate either a goddess or the embodiment of fertility. But research shows that, in contrast to the few unmistakable statuettes of women, there are hundreds of Paleolithic figurines from Western Asia and Europe that don't indicate gender, or seem to be of youths or slender girls. One paper of mine (Kehoe 1991) looks at this misguided interpretation from another angle. Carvings said to represent pairs of women's breasts ("mammiform"), with a bit of neck between, are actually oriented incorrectly in published pictures. Some from the Czech Republic were carved with a suspension ring in the back, and when the pendant is hung on a string through the ring, it clearly is a realistic carving of male genitals (Figure 6.2)! Very similar little pendants, mostly brass, were made by Romans; their soldiers liked to hang the miniatures on horses' harness and on tents over the door: here's a real man! The stereotype that "primitive people" fixate on fer-

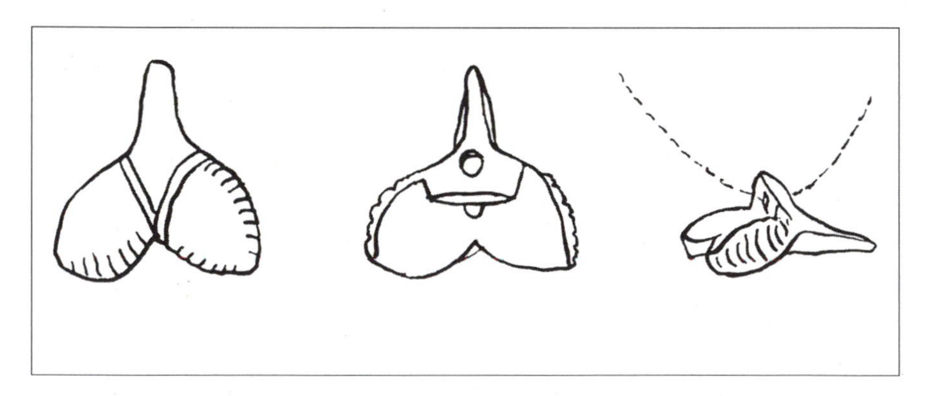

FIGURE 6.2 Small pendant representing male genitalia carved out of mammoth ivory, Dolní Vestonice, Czech Republic, ca. 23,000 B.C.E. (front and back, and as it hangs from string through carved suspension hole). *Drawing by Alice B. Kehoe.*

tility symbolized by women doesn't hold when we survey the variety of Paleolithic and Neolithic art and design. Nor should we deny the possibility that ancient people, like Roman soldiers, were having a bit of fun with good-luck amulets.

There seems to be no archaeological support for a peaceful, gender-equal world in the distant past. Nonetheless, the idea of such a world in the future appeals to many—women and men alike. As Cynthia Eller emphasizes, we don't need speculative interpretations of archaeological data to justify such an ideal: our common belief in the value of justice and human rights is a strong foundation for building a happier world.

■ ■ ■ SHAMANS

Another myth about "primitive peoples" is that all small societies, Paleolithic or contemporary non-Western, have "shamans" leading "shamanic religions." The myth was popularized in the 1960s by a self-announced historian of religions claiming, illogically, that there really is no history of religions but only the Sacred appearing again and again. Benefiting from the 1960s fad for mind-altering techniques and drugs, his book *Shamanism: Archaic Techniques of Ecstasy* became a classic, in spite of scholars detecting numerous inaccuracies. According to the book, non-Western peoples foster religious leaders who go into trance to experience the Sacred. The word "shaman," from the Tungus language in Siberia, had been familiar to Europeans since Russia conquered Siberia in the eighteenth century, and the book used the Tungus word indiscriminately to describe all ecstatic religious leaders, even for Paleolithic religions. This usage would make sense if one accepted the claim that all human religious experience is experience of an immanent eternal Sacred, but that is a *theological* proposition not amenable to scientific, archaeological, or historical proof.

Then, in the 1980s, David Lewis-Williams, a South African studying rock art, postulated that the country's rock art had been made by its Bushmen (Khoisan) natives, specifically by their "shamans" coming out of trance and depicting visions. Khoisan religious practitioners are not shamans, since, as we would expect given thousands of miles and great environmental differences, they have little in common with indigenous Siberian religious practices. Adding a new and even more reductionist note, the South African asserted that what all the "shamans" saw in their trances was a limited set of colored geometric figures produced by our hard-wired human brains. Instead of the Sacred manifesting to imperfect human intelligences, in his opinion these figures were simply a product of the genetics of the brain's visual cortex. Now all rock art could be easily, "scientifically," analyzed: from the Upper Paleolithic onward, nonliterate people mistook brain-produced patterns for spiritual enlightenment, were impelled to record them on rocks, and archaeologists need only discern the geometric figures, sometimes naively elaborated into pictures. The one-size-fits-all procedure *premising* "shamans" regularly going into trance is then read backward to affirm that we now have evidence (the rock art) of that universal type the shaman, in prehistory, throughout indigenous America, Australia, Asia, and Africa, wherever there is rock art. Critical review demonstrates, to the contrary, that both ethnographic understanding of religions among a large number of societies, and clinical knowledge of neural anatomy, genetics, and psychology of consciousness, amply indicate that the claim for universal "shamanic religion" is untenable; it cannot be used to interpret rock art or religions of the past. It doesn't even accurately describe the religious beliefs and practices of the indigenous Siberian nations who do call some members "shaman."

Real shamans—that is to say, these religious leaders of Evenki (formerly called Tungus), Sakhá, Khanty, and other "Small Nations," as Russians term the First Nations of their Siberian lands—are priests for their communities. They were trained by apprenticing to a practitioner, memorizing prayers and rituals and learning illusionist techniques to impress audiences that supernatural power had been summoned. During performances, drumming might induce shamans to lose consciousness of their body, and regaining consciousness, recount a magical out-of-body journey to abodes of spirits in order to retrieve sick persons' souls or to persuade game animal spirit herders to send a reindeer to the shaman's hunter clients. Other versions of shamans' supernatural journeys had them court a reindeer-spirit doe, win her as a wife, and oblige her family to send deer to their new son-in-law and his friends. The shaman vigorously danced the courting and ended by simulating nuptial intercourse with his invisible (to the audience) reindeer bride.

A French anthropologist witnessing such a ritual remarked that no one in trance could have performed such a narrative drama. Rituals of soul journeys or spirit possession seem to work by practitioners concentrating intently, to the point of losing everyday awareness of their bodies and ordinary sense perceptions. Drumming helps to disengage practitioners' minds. Feeling dissociated from one's body can also be an effect of nutritional deficiency, it can come when dancing to the point of hyperventilating, and it can be induced by a doctor or priest teaching a patient to concentrate fixedly on an anxiety or wish, a form of hypnosis. It is true that our human brains are genetically structured to be able to disengage exterior consciousness by intense focused concentration, so ritualistic use of this capacity is seen in many societies and may have been used by anatomically modern humans of the Upper Paleolithic. That does not specify Siberian shamans' beliefs or practices, developed in northern Asia among historic reindeer-hunters and their neighbors. Nor does it specify a three-stage trance that the South African rock art analyst supposed to be a universal human propensity; the trance experience he described results from taking certain hallucinogenic drugs or from hypnotically focused concentration.

We cannot use archaeology to verify Lewis-Williams's theory; it lies in the realm of belief, not science. But we can study other research questions on shamanism—for example, whether the practices and versions of this Siberian belief system spread across northern Eurasia, not only to northern Scandinavia where the linguistically related Saami ("Lapps") practiced it, but also into Alaska and across northern North America. Throughout the north, Inuit and neighboring American Indian religious leaders perform similarly to Siberian shamans in order to heal people and to persuade spirit beings to send animals for their communities' hunters. Alaskan Inuit and Siberian Chukchi, on either side of the Bering Strait, traded with each other for centuries and occasionally married. Each pursued trade deep into the interior of their continents, seen archaeologically by finding Asian iron tools in northern American sites. Siberian shamanic techniques occur throughout the northern half of North America, from the Arctic and Subarctic (Canada), as far south as the northern U.S. Plains in the Lakota *yuwipi* ritual. South of the Dakotas and Midwest, religious healers and priests' practices are not so similar to those of Siberia, and most of South America lacks the shaman's essential instrument, a hand-held drum. I myself have watched Blackfoot elders in northern Montana instruct a young man of their community in using the drum, held in the left hand above one's heart to mimic the heartbeat. Religious practices that don't contain this type of drum should not be termed "shamanic."

Inference to the best explanation can be found elsewhere in rock art research. The Society for American Archaeology selected Kelley Hays-Gilpin's *Ambiguous Images* for its 2005 award for best archaeology book for a general audience. Hays-Gilpin's title says it all: most rock art images are ambiguous. She makes the point that Siberian rock art can be placed into a time series, from around 4000 B.C.E., through the succeeding Bronze Age and Iron Age, through the Scythians described by Classical Greek geographers, and on to fully historic medieval records. A "deer" (elk, reindeer) deity or spirit is pictured on rocks and ornaments for millennia, but after all, deer/elk/reindeer were, with fish, the staple foods of northern forest and tundra societies.. Both stags and does are pictured, sometimes with an anthropomorphic (human-like) lady who might be Mistress or Mother of all Deer, or perhaps the daughter of Deer Chief wooed by the human shaman for hunters' luck.

Yes, ambiguous. Hays-Gilpin shows, again and again, that inference to the best explanation (see Chapter 2) can require the archaeologist to stop short of one single best explanation to explain all phenomena. Her caution comes, in part, from her experience collaborating with First Nations people in the Southwest, listening to their explanations of rock art images and placement; she knows how often something that seems obvious to a Euroamerican can actually represent quite a different meaning to descendants of local indigenous communities.

Another archaeologist who studies rock art, Linéa Sundstrom, also regularly listens to indigenous neighbors of her sites, in her case Lakota (Sioux) in South Dakota. Like Hays-Gilpin, Sundstrom works backward in time from the twentieth century into pre-European eras. Sundstrom's thoughtful study of nineteenth-century accounts of Lakota beliefs and practices led her to recognize that certain rocks with many vertical cut and rubbed lines are mute records of Lakota women sharpening their bone sewing tools, choosing these landmark rocks to do so because legend said they could there invoke blessing on their work from a tutelary spirit. Women could envision new beautiful designs while they fasted, prayed, and polished their sewing awls at these rock shrines. Hays-Gilpin's and Sundstrom's work, avoiding grand generalizations, exemplify rock art research grounded in direct consultation with indigenous people.

Calling the religions of all non-Western small societies "shamanic" is, I believe, racist, because it lumps a huge diversity of cultural histories into one category that is contrasted with another set implicitly—and not so long ago openly—labeled "civilized." In this establishment of gross—and incorrect—generalizations, it is not unlike the 1960s "Man the Hunter" conference described in the previous chapter.

There seems to be something hypnotically seductive to Western academics in the idea that they have a skeleton key that allows them to understand the minds of premodern people everywhere: Paleolithic artists, Siberian hunters and herders, Kalahari Desert "Bushmen," Neolithic townspeople at Çatalhöyük, Latin American indigenous priests, native doctors in the Amazon forests or in Mayan jungle kingdoms, and North American First Nations religious leaders. Every society today is as evolved as every other, each has its own history, its own adaptation to historical, environmental, and political experiences. All societies evidence change through time. An astute British anthropologist commented that academics' search for "universal laws" of human behavior is "ersatz religion," a foolish faith. Several anthropologists who have worked with Khoisan communities, a neurophysiologist who researched mental dissociation, archaeologists who have studied African rock art, and others familiar with indigenous groups and their, or their forebears', rock art, have tried to point out flaws in the shamans-in-trance explanation, to little avail (Figures 6.3, 6.4).

Many so-called shamanic societies are actually conquered or colonized nations that formerly maintained formal priesthoods before being subject to

FIGURE 6.3 Figurine of a leader, Colima, West Mexico, ca. 100 B.C.E–A.D. 250. Clay, 19.5 inches high. Note the protruding "horn"—or hair bun?—above the forehead. *Amerindian Art Purchase Fund 1997-363, Art Institute of Chicago. © The Art Institute of Chicago.*

FIGURE 6.4 Gros Ventre leader Niatohsa, painted by Karl Bodmer in Montana, 1833. Niatohsa's protruding hair bun marks him as a man of power. Is this the correct interpretation of the Mexican hairdress? *Courtesy Joslyn Art Museum, Omaha, Nebraska.*

Western religious practices. And, ironically, some Western churches include practices that could be termed "shamanic" under this definition: religious leaders experiencing mystical oneness with Divinity, healing power, assertions that rituals bring a divine presence into the room or transform substances into a holy essence. "Accepting Jesus into one's heart" seems similar to "shamanic transforming," and we can see many thousands of paintings and sculptures inspired by fervent Christian belief. This is not meant to demean practices among Christian congregations, but to clear away stereotypes rooted in racism. Humans have evolved a wonderful capacity to feel transcendental experiences, which in multiple societies over thousands of years has led to a multiplicity of religions. Archaeologists trace out culture histories, and this basic procedure forbids our lumping numbers of disparate societies into simplistic categories and so denying their histories.

Like Marija Gimbutas's myth of goddess-ruled societies, the idea of shamanism picked up popularity with the general public during the New Age era of Western culture. The emphasis on hallucinogenic drugs, on altered states of consciousness, on a simpler, better past, all appealed to the culture of the 1960s and 1970s. It is a frequent pattern to turn a wished-for present into a mythical past, but no good fairy, or goddess, can grant us archaeological evidence just for a wish.

ARCHAEOLOGY OF THE BIBLE

Archaeological research has also been called upon to illuminate mainstream Western religions. Much of both the Old and New Testaments of the Bible appears to recount historical events. Even if one discounts miraculous details, such as Joshua's trumpet call causing the walls of Jericho to collapse, it seems reasonable that an attacking Israelite army did at some time breach the city walls, and that archaeological excavations could reveal and date a layer of collapsed stones and bricks, weapon fragments, and perhaps soldiers' bodies. As archaeology became a professional discipline using scientific methods, during the second half of the nineteenth century, many Christians and Jews wanted to use it to identify places significant in the holy books. They assumed the Bible was basically a historical chronicle in which divine messages were embedded.

Back in the mid-nineteenth century, the Bible came under attack, for the first time, on several fronts. Most directly, "higher criticism" in universities revealed the biblical text as a complex series of stories, documents, and principles that were written, edited, and codified over the span of a thousand years, not a seamless document revealed directly by God. More to our point, explorations both in Western Asia and Egypt, and in Paleolithic sites in Europe, brought to light not only expected ruins at likely biblical sites, but unguessed-at cities and histories in biblical lands, and astounding evidence of humanity's great antiquity. Paleolithic artifacts and an association of humans with Ice Age animals stopped being interpreted as evidence of the biblical Great Flood. Rather, geological studies showed that sediments accumulated slowly over eons of time and that human remains and artifacts within them were equally old. Rejecting a literal reading of Genesis picturing God constructing our world in six days, some nineteenth-century thinkers rejected the idea that every kind of plant and animal is the way God made it.

Scottish journalist Robert Chambers created a scandal with his 1844 *Vestiges of the Natural History of Creation*, published anonymously, portraying a kind of Big Bang nebula from which atomic particles coalesced into chemical elements that gradually evolved into stars and planets and then into organisms or inorganic forms such as rocks. Chambers's world would be continually changing, continually producing new, more complex or successful creatures. His political message was unmistakable to his Victorian readers: English society, too, was inexorably changing. This scandalized England but helped, as scandals often do, to sell the book for another fifty years. And English society did change substantially during that half-century, discombobulating plenty of formerly well-off aristocrats and gentry. Among the latter was Charles Darwin, whose inherited income allowed him to devote himself to meticulous scientific studies of several kinds of animals. Darwin saw that Chambers might be on the right track, but Chambers's popular journalistic style, political overtones, and insisted-upon anonymity, put off serious readers. Fifteen years after Chambers's *Vestiges*, Darwin was pushed to publish his own solidly supported hypothesis of descent with modification accounting for the myriad of organisms we see. His theory that natural selection influences species modification stands today as the foundation of biological sciences, an outstanding inference to the best explanation.

The weight of geologists' argument that rock strata needed millions of years to accumulate convinced most people that Genesis in the Bible had to be read as poetry and metaphor, not literally—although there are still fundamentalist denominations insisting that their favored translation of the Bible is absolutely word for word God's truth, never mind contradictions within the biblical text itself.[2] For scholars of religion, contradictions can be explained through "form criticism," separating out different passages according to linguistic and contextual clues, organizing them into several strands of writing, and attempting to date each strand based upon internal clues. For more than a century now, scholars hoped that the cities buried under sand and soil in Palestine would furnish archaeological proof for events in the Bible and data to assist in deciding historians' debates. Believers sponsored archaeological projects to link specific ruins to biblical personages and activities, while archaeologists struggled to interpret their data scientifically, often at odds with the biblical story. As a result, instead of settling debates, archaeology has fueled them.

Biblical archaeology began in 1838 with surveys in Palestine and Syria by a faculty member of Union Theological Seminary in New York. It took off around 1880 with big excavations at cities known from the Bible, tied into business and political manipulations by the ambitious powers of the time, Britain, France,

---

[2] See, for example, the two separate stories of the creation of humans: Genesis 1:27, "And God created man in His own image, in the image of God created He him; male and female created He them." Then, Genesis 2:22, "And the rib, which Lord God had taken from the man, made He a woman, and brought her unto the man."

Germany, Italy, and after 1900, the United States. Diplomats from these powers negotiated concessions from Arab countries giving their teams exclusive permits to excavate, with artifacts to be divided between the host country and the team's national museum in Europe or America. Egypt as well as the Near East was coveted by biblical archaeologists because it figures in several episodes of the Old Testament (the stories of Moses and of Joseph and his brothers). For several decades of the middle twentieth century, American archaeology in Bible lands was dominated by William F. Albright, a scholar teaching at Johns Hopkins University. His many, well-written books seemed to document a great deal of Bible history—for example, identifying the city of Jericho and fallen walls presumably brought down by Joshua's trumpets (Joshua 6:20). Joshua took over (at God's command) upon the death of Moses (Joshua 1:1) and led the invasion of what is now Israel; the Jordan River parted, like the Red Sea did for Moses, for the Chosen People to cross dry-shod. But when examined by more secular archaeologists, such as England's Kathleen Kenyon, who excavated Jericho in the 1960s, the identification fell short. What Kenyon shows at the oasis of Jericho is a deep series of occupations going from earliest farmers at the end of the Pleistocene Ice Age, through thousands of years as a commercial town. Not only were there no inscriptions naming a leader "Moses" or "Joshua", but it is likely that the city was uninhabited at the time that scholars think Joshua existed—if he existed at all. The same situation was demonstrated at the many other sites simplistically associated with biblical personages and events by eager theologians.

Serious archaeology in "the Holy Land" is now labeled "Syro-Palestinian archaeology," the label indicating its practitioners are fitting its archaeological record into world prehistory and secular history, rather than attempting to confirm biblical texts. The more archaeological data are gathered from surveys and excavations, the more complicated the region's history seems to become, especially to the large audience curious about how the archaeological record fits the Bible. Data from the areas believed to be the Bible's Israel and Judah, from late second millennium B.C.E. to Assyria's conquest in 721 B.C.E., indicate there was never, in this period, a "United Monarchy" of the two Jewish kingdoms. There is no evidence outside the Bible of famous biblical kings named Saul and Solomon; evidence of someone named King David only surfaced in the 1990s. These biblical heroes seem to be amalgams of real and legendary leaders. The "Babylonian exile" lamented by Ezekiel and Jeremiah after the destruction of Jerusalem in 586 B.C.E. is equally suspect. Archaeology shows a continuing population of farmers and herders, and their idols (as the Bible calls their images of deities), from the end of the Pleistocene into the present. Some leaders were exiled, their names appearing in Babylonian documents. But new ideas, technologies, and people kept on coming into the land, styles changed, national histories and religions were reworked. When (in Ezekiel 16:26–29) the Lord chastises "Jerusalem" for "prostituting herself" to Egypt, Assyria, the Philistines, and Chaldea, we have a truer picture of the little Jewish kingdoms' insecure political existence.

At the crux of the controversy over biblical archaeology is the belief held by fundamentalists that the truth of their faith depends upon whether the Bible is reliable history. It all leads up to whether Jesus of Nazareth, a historical person called the Christ, was in fact bodily resurrected after his death. For most people, rubble walls at Jericho have no connection to this miracle. Certainly, few scientifically trained archaeologists wish to venture opinions on matters of religious faith. Bible literalists have turned their attention more to signs of impending Armageddon (named after the archaeological site of Megiddo in central Israel), the prophesied final battle between good and evil. (Some are putting money into a project to bring red cattle to Israel from the United States, based on a passage in Numbers 19, where the Lord instructed Moses and Aaron to sacrifice an unblemished red heifer, which these Americans plan to do at a new Temple in Jerusalem, bringing about the End of Days for earthlings.) The greater number of Christians, Jews, and Muslims welcome the illumination archaeology brings to the cultures of the eastern Mediterranean within which the prophets of all three religions lived. We archaeologists can tell you what tombs looked like that the body of Jesus of Nazareth would have been put into when Rome ruled Judea, but we can't verify miracles—that's not our business.

A Case in Point: Yahweh's Canaanite Goddess Consort

William Dever, who calls himself a Syro-Palestinian archaeologist (in contradistinction to conservative biblical archaeologists in the same countries), brought together finds from his own and others' archaeological excavations in Israel that show many local and household shrines invoking the deities Ba'al and Asherah as well as the Jewish God, Yahweh. The phrasing of one Hebrew inscription on a painted jar from eighth-century B.C.E. Kuntillet 'Ajrûd, a travelers' stop in the eastern Sinai desert region, speaks of Yahweh and "his" Asherah, a Canaanite goddess and possibly his consort (Figure 6.5). The lady is often depicted with lions—not, it seems, the historic Lion of Judah, but symbols of her own power, the power of reproduction and growth (might she be the same Lady as the figurine with her hands on felines' heads at Çatalhöyük?). Kuntillet 'Ajrûd was a fortified hostel serving a major caravan route, so perhaps the Egyptian and Phoenician symbols and deities indicate the shrine building could serve traders from any of the nations using the route. Dever thinks that the number of sites from the early part of the first millennium B.C.E. with pictures of the goddess or her name indicates that the Israelites amalgamated deities from the Canaanite communities they battled or settled among, and from adjacent Egypt, Phoenicia, and Mesopotamia, the same powerful neighbors denounced by Ezekiel. Ordinary Israelites of that time, it seems, were friendlier to their neighbors' gods than the Old Testament wanted to acknowledge.

The Old Testament books of Joshua, Judges, and Samuel contain accounts of kings of Israel forbidding the worship of these deities and destroying their shrines at various locations. Some of the archaeological sites with toppled and

FIGURE 6.5 Yahweh's consort, the Canaanite goddess Asherah (*upper right*), playing her lyre for the dwarf gods Bes, patrons of music and dancing. Scene painted on a storage jar. *Courtesy Zev Meshel, from Meshel 1979: fig. 12.*

broken-up altars and figurines may be among those referred to in these books of the Bible. If so, archaeology has confirmed the actuality of some events chronicled, although because we lack inscriptions at the sites, the evidence is not conclusive. At the same time, the discoveries do show us the persistence of a folk religion worshipping both male and female deities in addition to the Bible's Yahweh, even pairing him with the principal goddess as consort. Not until the defeat of the kingdom of Israel and the Babylonian exile of a segment of its population, in 586–538 B.C.E., did Jewish leaders succeed in purging these deities from association with Yahweh. Dever's reliance on archaeology to illuminate Israelite society and history complements theologians' and historians' close reading of biblical texts, creating a more complex and nuanced history than either alone will yield.

Book of Mormon

Less well known than biblical archaeology are the efforts of members of the Church of Jesus Christ of Latter-Day Saints (Mormons) to identify places in the Americas with those mentioned in the Book of Mormon, the Church's central scripture. Joseph Smith, the martyred founder of the Church, announced in 1830 that a divine messenger had presented him with the book in an ancient language, and he had translated it. His followers accept the Judaeo-Christian Bible along with Smith's scripture describing a migration out of Judah, in 600 B.C.E., across an ocean to inhabited lands that contemporary Mormon scholars believe was Mesoamerica (Mexico and Central America). Jesus Christ is said to have traveled to and preached in this remote land. The Church teaches that the Book of Mormon was written about A.D. 400 by a man named Mormon, the last in the royal dynasty founded in America by Nephi. His history mentions other groups or nations, especially Lamanites and Jaredites, allied with or combating the Nephites. In the 1950s, Mormon researchers began trying to match passages in the Book of Mormon with places, peoples, and cultural phenomena in prehistoric Mesoamerica, and to sponsor direct archaeological survey and excavation. Their work is intended to amplify the Book of Mormon rather than test the authenticity of their scripture—that is to say, the Mormon projects are planned with an understanding of what archaeological data can, and cannot, evidence.

Latter-Day Saints archaeologists present their research data in secular, professional archaeological meetings, saying nothing about Book of Mormon interpretations published separately for Church readers. Where biblical archaeologists used to clash with secular historians and archaeologists over details within historical eras that no one doubts did exist, Latter-Day Saints archaeologists have to deal with colleagues who consider the Book of Mormon a nineteenth-century fabrication, and therefore of no possible value in interpreting Mesoamerican prehistory. To a non-Mormon archaeologist, the books published in Utah for Latter-Day Saints read like science fiction, strongly grounded in good data but leaping off into a narrative few outside the Church have heard of.

Affiliation to a church, whether Mormon or more conventional Judaeo-Christian, raises suspicion that hope or expectation of matching scripture affects inference to the best explanation. With Latter-Day Saints, their acceptance of the premise that Nephi and his descendants were in Mesoamerica is so far from mainstream archaeological knowledge that it vividly illustrates how important it is to recognize underlying premises in a scientific argument. Working from acceptance of the Book of Mormon, Church members have been compiling similarities between "Old World" and "New World" pre-Columbian cultures, expecting that the weight of numbers of parallels may tip the balance toward general affirmation of their scripture's validity. Their exhaustive bibliographies present much intriguing, and some convincing (to some of us secular archaeologists) evidence for transoceanic contacts between the Americas and other parts of the world before 1492, yet very little in these lists would point to the Book of Mormon explanation of a migration from Judah under Nephi. That would take discovery, in undisturbed sealed occupation remains, of *texts* naming Nephi or other figures from the scripture; material culture similarities are not sufficient to link a site to named individuals. Historical archaeologists have found over and over that nailing data from archaeology to documented individuals can be difficult, even when you're dealing with, say, the foundations of a well-known nineteenth-century brothel near the Capitol in Washington and wondering whether jewelry recovered was owned by the madam. As explained in Chapter 2, the *chain of signification* from data uncovered to interpretation within a historical framework is likely to have many steps, and each step should be supported by evidence. From an archaeologist's point of view, the Book of Mormon is an important religious document for a contemporary North American religious sect, but not a source on the archaeology of Central America.

■ ■ ■ IS ARCHAEOLOGY CAUGHT IN A CONFLICT BETWEEN SCIENCE AND RELIGION?

Most people will say that science and religion conflict. Science is atheistic. "*A*-theistic," *without* a god. Yes, insofar as science is *by definition* restricted to observing, analyzing, and interpreting the natural world. The *super*natural is excluded from the domain of science; it is not within its *jurisdiction* (think of courts of law, setting out the boundaries of a court's jurisdiction). There is no conflict. Any sensible person can see that our human capabilities are not omniscient (all-knowing) or omnipotent (all-powerful). For that very reason, we observe our world carefully, try to be precise and detailed in sharing knowledge we have gained, distinguish between what can be observed directly by any interested person—*science*—and what is said to have been revealed by supernatural authority—*religion*. Ballyhooed "wars between science and religion" are political ploys, propaganda designed to discredit an opponent.

Archaeology on the whole offers little support for religious claims. Scientific archaeology doesn't play in the same ballpark as religion. From another perspective, archaeology has greatly enhanced histories of religions and understandings of religions in societies. Archaeology gives material shape to verbal images in religious texts. It is somewhat arbitrary to disentangle "religion" from politics, economics, literature, and theatrical performances. Many data are ambiguous for an archaeologist hoping to recapture the meaning they held for the people who made and used them (as Kelley Hays-Gilpin emphasized). Given limitations imposed by the archaeological record, no more than the nonperishable residue of once-living societies, that record does help us visualize material dimensions of human lives. The domain of religions intersects those dimensions, but its core is independent of archaeology.

Sources

Dever, William G., 1990. *Recent Archaeological Discoveries and Biblical Research*, Seattle: University of Washington Press. See pp. 119–166 on the Israelite cult.

—. 1999. American Palestinian and Biblical Archaeology. In *Assembling the Past*, edited by Alice Beck Kehoe and Mary Beth Emmerichs, pp. 91–102. Albuquerque: University of New Mexico Press.

Eller, Cynthia, 2000. *The Myth of Matriarchal Prehistory: Why an Invented Past Won't Give Women a Future*. Boston: Beacon Press.

Elster, Ernestine S., 2007. Marija Gimbutas: Setting the Agenda. In *Archaeology and Women*, edited by Sue Hamilton, Ruth D. Whitehouse, and Katherine I. Wright, pp. 83–120. Walnut Creek, CA: Left Coast Press.

Evans-Pritchard, E. E., 1963. *Essays in Social Anthropology*, Glencoe, IL: Free Press.

Finkelstein, I., and Neal A. Silberman, 2006. *David and Solomon*. New York: Free Press.

Harvey, Graham, editor, 2003. *Shamanism: A Reader*. London: Routledge.

Hays-Gilpin, Kelley, 2004. *Ambiguous Images: Gender and Rock Art*. Walnut Creek, CA: AltaMira.

Helvenston, Patricia, and Paul G. Bahn, 2006. Archaeology or Mythology? The "Three Stages of Trance" Model and South African Rock Art. *Cahiers de l'AARS* 10:111–126.

Hodder, Ian, 2006. *The Leopard's Tale: Revealing the Mysteries of Çatalhöyük*. London: Thames and Hudson.

Kehoe, Alice Beck, 1991b. No Possible, Probable Shadow of Doubt. *Antiquity* 65 (246):129–131.

—. 1996. On an Unambiguous Upper Paleolithic Carved Male. *Current Anthropology* 37 (4):665.

—. 2000b. *Shamans and Religion: An Anthropological Exploration in Critical Thinking*. Prospect Heights, IL: Waveland Press.

Meshel, Zev, 1979. Did Yahweh Have a Consort? The New Religious Inscriptions from Sinai. *Biblical Archaeology Review* 5(2):24–34.

Sidky, Houmaym, 2007. Haunted by the Archaic Shaman: Himalayan Jhăkris and the Discourse on Shamanism. Ms. under submission to publisher.

Sorenson, John L., 1998. *Images of Ancient America: Visualizing Book of Mormon Life*. Provo, UT: Research Press of the Foundation for Ancient Research and Mormon Studies.

Sundstrom, Linéa, 2004. *Storied Stone: Indian Rock Art of the Black Hills Country*. Norman: University of Oklahoma Press.

Thompson, Thomas L., 1999. *The Mythic Past: Biblical Archaeology and the Myth of Israel*. New York: Basic Books.

7 "Diffusion" versus Independent Invention

Dogmatic orthodoxy versus common sense is very evident in controversies over whether people crossed the oceans before Columbus in A.D. 1492. Common sense points to the indisputable fact that men and women crossed open ocean to get to Australia fifty thousand years ago in the Pleistocene. There is no other way that continent could have been populated. Polynesians sailed to hundreds of islands in the Pacific many centuries before Europe's Age of Exploration began in the fifteenth century. Our own experience and common sense tell us how commonplace it is to get a new idea or technology from someone in another culture, whether it's pizza and tacos, or yoga, or a Japanese video game. Conversely, it is rare to independently invent something—if you try to get a patent on an invention, you'll find it isn't easy to prove originality. Nevertheless, it has been conventional in American archaeology to assert that the Americas were isolated from contacts with other continents after initial population migrations over the Bering Strait into Alaska around ten thousand years ago, and that consequently, all the native societies of the Americas independently invented everything in their cultures other than Ice Age hunters' equipment. In this chapter, we look at boat technology and at evidence, some beyond doubt and much controversial, for intersocietal contacts before the European Age of Exploration and colonization. While many of my colleagues don't agree with all the claims I make in this chapter, I believe that the model of intermittent contacts best explains the cited features of indigenous American cultures.

■ ■ ■ POLYNESIANS IN THE AMERICAS

Easter Island lies more than 1,400 miles (over 2,200 kilometers) across open ocean from the nearest Central Polynesian islands. The island itself is a few square miles smaller than the city of Washington, DC. On maps of the Pacific, it is a tiny speck. About a thousand years ago, Polynesian explorers sailing in large canoe-hulled ships found this small island; sailed back to Central Polynesia and loaded ships with men and women colonizers, crop plants, food and water for the long voyage and to feed them until their first harvests; and sailed this convoy to the island they named Rapa Nui, where descendants of the Polynesian settlers still live. During the centuries when Europeans were bogged down raiding and battling between their competing kingdoms, Polynesian explorers discovered the Hawaiian chain of islands away to the northeast, and New Zealand to the southwest, both farther away than Rapa Nui, although much larger. In each case, the explorers navigated back to their home islands in the central Pacific and outfitted colonies. Now, think about this: if for centuries, Polynesians deliberately explored the entire Pacific Ocean and had the technology to navigate precisely and transport settlers with their tools and foodstuffs, does it seem likely that in all those centuries, not one Polynesian ever discovered America?

Clearly, there is a strong probability that Polynesians touched the American continents. Did they colonize? Apparently not; they were looking for *uninhabited* lands for their expanding populations, and they would have quickly seen that the American continents were already well populated. Did they do a bit of trading? Perhaps a small party of men took up residence, marrying locally, reconnoitering for transportable resources. This could have happened, leaving no imperishable evidence for archaeology. One possible trace of such a minor venture may be the unusual sewn-plank-built canoes of the Chumash and Gabrielino (Kumivit, in their own language) of the southern California coast, and the names these nations used for the canoes, *tomolo* and *ti'at*. Except for one other sewn-plank canoe occurrence on the coast of Chile, indigenous American boats were rafts, log dugouts, lashed reed bundles, or frames covered with bark or hide. Radiocarbon-dated canoe planks from Chumash territory indicate the sewn-plank boat was in use in the mid-first millennium A.D. Two-piece bone fishhooks similar to ones made by Polynesians of the time also suddenly appear on the California coast in mid-first millennium A.D. The Chumash language seems unrelated to any other—indigenous American or outside the Americas—except for the word *tomolo*, which doesn't fit Chumash word structure but is plausibly derived from first-millennium A.D. Hawaiian Polynesian words *tumu raakau*.[1] The Gabrielino, who moved to the coast relatively recently before European contact, called their sewn-plank canoes *ti'at,* and boats in general, *tarayna*. As with Chumash *tomolo*, the two Gabrielino words are strange for that language (which is Uto-Aztecan, related to Ute, Shoshone, Hopi, and Mexican Nahuatl), but can be derived from early Hawaiian words for sewing with a small wooden implement *(tia)*, and for hewing or carving wood (*taraina*). Derivation of the indigenous Chilean word for the sewn-plank canoe there, *dalca*, has not been studied.

Many archaeologists are bemused by linguists' multi-step identifications of historic words with cognates in related languages and proto forms in the common ancestor of the related languages. It is easy to scoff that *tomolo* and *ti'at* aren't complicated words, and that sewing planks to build up a canoe can be logically developed from making dugouts and sewing. Yet why were these few loan words and technologies for fishing and sailing held in common between southern California and the Pacific islands? It is not a stretch to argue that this was because these were what the Polynesians traded when they landed in California.

Credibility for the California Chumash–eastern Polynesian contact is heightened by the incontrovertible fact that sweet potatoes, *Ipomoea batatas*, were grown in both South America and Polynesia before European explorations in the sixteenth century, and in both regions were called *kumara*. The plant is native to South America, and its spread through the Pacific around a thousand years ago is documented with archaeological finds. There is no way sweet potatoes spread through the Pacific except by being carried in boats, and sharing the same word for the plant implies peaceful trading contact ("Me Tarzan. You Jane. That," pointing, "kumara."). Another biological organism, bones of Polynesian chickens that could only have been carried by people in boats, turned up in excavations of a pre-Columbian site in Chile, strengthening the probability that plank-built *dalca* boats there were copied from Polynesian travelers.

Given Polynesians' indisputable seafaring skills and drive to explore, plus the presence of cultivated sweet potatoes called *kumara* in America and Polynesia before European contacts, why do many archaeologists shy away from discussing possible trans-Pacific contacts before Magellan's voyage, in 1520–21? They will quickly answer, shuddering, "Diffusion." A century ago, an Australian-born anatomist, G. Elliot Smith, tried to persuade readers that all civilizations emanated from ancient Egypt. Egyptology being as popular then as now, Smith's belief attracted enthusiasts and still persists, alongside others' conviction that drowned

[1] I am simplifying the linguistic derivation details. See the Jones and Klar paper for the full argument.

Atlantis was the mother of all civilizations (see Chapter 3). Somewhat more limited "diffusionism" would have all American civilizations founded by White Gods, or legendary Lost Tribes of Israel (similar to the Mormons' tribe led by Nephi; see Chapter 6), or North African Carthaginians after their defeat by Rome, or rulers from south India fleeing Aryan conquerors (Chapter 3). Characteristic of such enthusiasts is the energetic aggregation of pictures and historical facts, frequently ignoring context—that *sine qua non* ("that without which, nothing") of archaeologists. When archaeologists dismiss such accumulation of alleged evidence, enthusiasts may see deliberate suppression of challenging points of view, even conspiracies to hide secret knowledge.

The late Barry Fell, a marine biologist and self-taught epigrapher (researcher studying inscriptions), is a prime example of these enthusiasts. He wrote that publicizing his ideas became "something of a ministry." He claimed to have deciphered hundreds of pictures, inscriptions, and scratches on American rocks as evidence of expeditions by ancient Old World nations speaking forgotten languages. To do so, he drew from dozens of languages and archaeological evidence from many continents, but not in the consistent, limited, rigorous way that the proponents of the Chumash–Polynesian hypothesis did; nor did he present indisputable biological identifications such as sweet potatoes and Polynesian chickens. Fell's followers thrilled to his translations supposedly of esoteric prayers and invocations. He ended his first book,

> So to all these modern descendants of ancient heroes and heroines who crossed the great waters we send greetings, *ceud mille faillte*, ten thousand greetings from Iargalon, the "Land-that-lies-beyond-the-sunset." May the eye of Bel look kindly upon us all. (Fell 1976:294)

This language reminds me of Tolkien's *Lords of the Ring*, the fantasy books and movies about little hobbit people journeying far, meeting elves and wizards, on a quest to recover a magic ring. And that is how archaeologists and linguists view his work, as romantic fantasy.

■ ■ ■ BEHIND THE DISMISSAL OF "DIFFUSION"

Between the diffusion enthusiasts and mainstream archaeologists disdaining any pre-Columbian transoceanic contacts (including between Latin America and Anglo America!) are a small number of scientifically trained avocational researchers who drew away from the fervor of Barry Fell and his ilk, plus reputable archaeologists and geographers who acknowledge the high probability that Polynesians and other seafarers reached America before 1492. The evidence is multitudinous, far too much to list.[2] Here, we must discuss reasons for the conventional dismissal of pre-Columbian transoceanic contacts.

Nineteenth-century anthropologists and archaeologists generally believed that European societies had achieved a more complex, "higher" civilization than any non-European cultures. Most took it for granted that their "race," Europeans, was more evolved away from ancestral primates than are peoples of other continents. Skin color was the prime criterion of evolutionary progress, so the whitest people, sun-deprived northern Europeans, were at the "higher" end of the scale, "yellow" northern Asians below them, "brown" and "copper-colored" southern Asians and American Indians next below, and "black" Africans, western Pacific islanders, and Australians at the bottom. That skin color correlates with closeness, or distance from, the Equator with its strong sunlight is obvious but wasn't a welcome explanation; White nations of northern Europe wanted resources from warmer regions, and an ideology of evolutionary progress visible as whiteness legitimated conquering them. Invasion and conquest were touted as a noble sacrifice, the "white man's burden" of "civilizing our little brown brothers." In the United States, propaganda claimed it was America's "Manifest Destiny." Two corollaries to the ideology were (1) that "colored races" had simple societies unchanged over millennia ("Stone Age people," "primitives"), and (2) that civilization had to be taken to them by more evolved nations.[3] Reports of intelligence, ingenuity, or inventions of conquered "people of color" might be dismissed as exaggeration or foolish sentimentality.

By no coincidence, this grossly racist ideology waned as European and Euroamerican empires found it too costly to continue the "white man's burden" of administering colonies. At the same time, industrial economies in the imperial nations shifted to employing more of their citizens in marketing, finance, and other late-capitalism white-collar jobs, moving industrial production to cheaper labor overseas and relying on immigrant "people of color" to perform menial jobs. When this trend got to the point of outsourcing computer engineering to a former colony, India, the old ideology began to shake.

What does this have to do with archaeology? In 1986, a regular international meeting of archaeologists scheduled for England was jeopardized by members

[2] See publications by Stephen C. Jett, a historical geographer who has conscientiously and carefully compiled voluminous data on pre-Columbian transoceanic contacts.

[3] See Chapter 5. All human populations are equally evolved, that is, descended through generations from earlier humans. Differences between populations are primarily adaptations to the regions they inhabit, as in the example of skin color correlated with sunlight strength. Societal differences, such as having a system of writing, correlate with potential for trade linked with location, climate, and resources.

campaigning to include colleagues "of color," native-born archaeologists from former colonies ("Third World countries"). The international organization split, with the challengers forming a World Archaeological Congress (WAC) dedicated to respecting cultural diversity and facilitating attendance of colleagues from less wealthy countries.

WAC is overtly political, asserting that none of us can be "above" or "outside" politics: choosing not to vote or to be active in "politics" is itself a political choice with repercussions on our community. Efforts to assist people from poorer nations to obtain advanced professional training and participate in leading archaeological projects, designing projects to elucidate questions of particular interest to these colleagues (for example, finding evidence of Hopi migrations; see Chapter 4), assisting in repatriating objects of value to indigenous communities, and working with indigenous groups toward social justice are among goals espoused by WAC. As non-Western archaeologists are more integrated into mainstream archaeology, and our larger society more forcefully repudiates racism, some of the assumptions that undergirded nineteenth-century diffusionist theories *and* the reaction against them by twentieth-century archaeologists are weakening.

With the twenty-first century, more sophisticated theorizing has superseded "superior race" beliefs and naive faith in independent inventions. A bold example by two European archaeologists argues, with much detail on sites and artifacts, that in the second millennium B.C.E., ambitious northern European leaders traveled to Mediterranean cities to learn of the world and buy fine ornaments and weapons, returning with religious as well as lifestyle ideas and inviting craftsmen and traders to their homelands. Through these journeys, Greek bronze swords, a gold-foil diadem crown, and gold foil to wrap around sacrificed cattle's horns were carried to northern Europe, eventually to be excavated by archaeologists there. These intersocietal contacts not only left direct copies of Mediterranean symbols, but also stimulated innovations. Independent invention of similar cultural traits is actually very hard to prove, while intersocietal contacts can be demonstrated historically to be frequent sources of cultural change. Archaeologists need to realize that before modern mechanical transportation, people *expected* to take months or years for a journey, spending lengths of time in foreign parts, and though the time for journeys was longer than we would afford, curiosity about distant places and desire for mementos were as strong as they are among us today. It is ironic to see that thirty-five hundred years ago, it was northern Europeans who admired "high civilization" among nations to the south. Cultures change, and so do the reputations of people once considered the "most cool," as we might say.

Science and Intersocietal Contacts

Joseph Needham (1900–1995) was an English biochemist who had been impressed with Chinese scientists invited to work in his pioneering laboratory during the 1930s. When he retired at age sixty-five from directing his laboratory, Needham started researching a question that had intrigued him for decades: why, given the sophisticated science and engineering developed in China over three millennia, did the country never break through with a scientific revolution like that in the West in the seventeenth century? What he expected to be a historical essay ready after a year or so of reading and analyses turned out to require a series of thick volumes, *Science and Civilization in China*, collaboration with hundreds of scholars throughout the West and Asia, and Needham's startling conclusion that Chinese science had been superior to that in the West consistently until about the eighteenth century. China didn't have a scientific revolution because it had fostered good, practical science as a matter of good government. As so often happens with research, finding an answer to the initial question stimulated Joseph Needham to extend his pursuit of information. The spread of a number of Chinese inventions, such as the magnetic compass (lodestone), failure of other inventions to go beyond China (for example, the flat-bottomed Chinese sailing ship, the *junk*, a safer and more commodious cargo vessel than Western sailing ships), and particularly inventions that spread widely in spite of being closely guarded secrets, such as gunpowder, fascinated Needham. Gunpowder and guns were the nuclear weapons of the late medieval period; anyone who betrayed the technology risked death, yet the technology and its terrible power could not be limited by government restraints.

The volumes of *Science and Civilization in China* detail the development and *diffusion* of many inventions and technical knowledge. Working through piles and piles of data, Needham and his principal partner in the enterprise, Lu Gwei-Djen (she had been one of the 1930s postdocs in his lab), set out basic principles for evaluating the likelihood that similarities are due to transmission from an originating group to a recipient. One criterion is *collocation*, the number of items or traits together (collated) in the feature; the more complex the artifact or steps in its manufacture, or the larger the number of items that occur together, the higher the probability of transmission together. The other principal criterion is distance between originator and recipient, either distance in space or distance in time. Distance in space may be overcome by evidence of long-distance travel (ethnographic, historical, and literary accounts) or technology, such as ships or caravans. Distance in time, between postulated originator and recipient societies, may appear because of undiscovered or perishable data—as always in archaeology, *lack* of evidence does not *prove* a phenomenon never existed. Needham and Lu's principles are helpful guidance, rather than open-and-shut methodology.

The spread of printing in Europe in the fifteenth century is a good example of transmission. Originating in China, along with paper-making, the craft revolutionized information management in Europe when a German entrepreneur, Johannes Gutenberg, created movable metal alphabet letters to replace carved blocks like those used in China for printing its ideograph words. Gutenberg figured out that an initial investment in sturdy metal letters, frames to hold them, and a press would in the end pay off through making possible the (relatively) rapid reproduction of unlimited copies of a text. Another factor in Gutenberg's invention was paper, much cheaper, thinner, and easily run through a press than the other writing material then used in Europe—namely, parchment (made from animal hide). The obvious advantages of movable-type printing on paper stimulated many knock-offs by other businessmen, leaving a trail of books and records for the historian. The point for the archaeologist is that block printing and paper-making were common in China for centuries before the fifteenth, commercial contacts between China and Europe are documented in records and by artifacts especially from the thirteenth century on (by Marco Polo, for one), so clearly the *collocation* of printing texts with blocks (process #1) pressed on paper (process #2) originated in China not distant in time and linked across space with Europe by ships and overland caravans. *But* the transmitted basic processes were reworked in Europe, adapted to a very different form of writing. American anthropologist A. L. Kroeber, a predecessor of Needham, called such phenomena "stimulus diffusion."

Trying to cope with these complicated cases so typical of transmission discourages archaeologists hoping to discover general regularities of cause and effect. These archaeologists believe that for archaeology to be a science requires exactly measured data, formulation of hypotheses that can be tested against data, and generalizations about human behavior and societies derived by statistical assessments of correlations (Chapter 5, page 102). Archaeologists can't set up experiments using actual human societies—and ethics aside, the experimenter wouldn't live long enough to see the outcome—so they look for a "natural experiment," similarities in postulated cause factors in different regions or times (see footnote 1, Chapter 2). Those who distrust people's ability or willingness to acquire foreign cultural items hope to find similarities correlated with particular ecologies, population size, or technologies. It doesn't take much searching of the ethnographic literature to notice there are impressive differences between human responses to what must be significant causal factors: for example, to survive dry heat, Arab peoples in the African desert cover themselves in robes, whereas Numa in the American desert and Australian aborigines simply go practically naked. To survive Subarctic winter cold, northern Europeans and eastern North American Indians built insulated houses, while many Dené Indians of the western American Subarctic forests relied on fires at the front of open lean-tos. And while in northern Eurasia, reindeer are herded and some are harnessed to sleds, tethered for milking, or used as pack animals, in North America closely related caribou are hunted as wild game.

What accounts for the difference between Eurasian and American usage of *Rangifer*, the genus of both reindeer and caribou? History. Northern Eurasians were in contact with societies that herded, milked, and harnessed cattle, horses, donkeys, goats, camels, water buffaloes, or elephants. The idea of taming another large mammal was transmitted, many times over. Northern American peoples did not see anyone taming a large mammal; urban societies south of them did quite well with water transport and human porters. Only Iñupiaq (Eskimo) in westernmost Alaska saw tamed reindeer on their visits to Chukchi across the Bering Strait, and about a thousand years ago, "stimulus diffusion" inspired Iñupiaq ancestors to harness dogs to sleds. Whether it was further contact, between Arctic dogsled users and ancestors of northern Plains Indians, that inspired Plains people to breed big dogs to carry packs and drag travois sleds, we do not know. Long-distance travel lures many humans, if the person doesn't have to be back at work on Monday, and our species survived and prospered through ingenuity in adapting raw materials to fulfill needs.

"Diffusion" or Psychic Unity?

A note here: I put "diffusion" in quotation marks because most discussion among archaeologists about "diffusion" is about intersocietal contacts. The word "diffusion" comes from the Latin *fundere*, to pour, and means to spread out, as in pouring out, or to intermingle. It is correctly used to describe molecules spreading through a gas or the way telephones spread throughout a country, as chronicled in classic sociological studies by Hägerstrand (1967) and Rogers (1962). An archaeologist can speak of a pottery style diffusing through a society, or a stone blade hafting style diffusing through North America. Diffusion may result from the introduction of a style, symbol, or technology, but not inevitably—the introduced trait could become limited to just a few people, or never accepted. I had a boss who grew up eating hominy stewed with pigs' feet and introduced the dish to us many times at get-togethers, without it ever diffusing among his employees. Polynesians likely introduced sewn-plank-built canoes to ancestral Chumash in California, but the technology did not diffuse beyond coastal southern California.

James Ford, an archaeologist considered by his contemporaries to be an outstanding theoretician, compiled a series of charts listing early appearances of ceramics, platform mounds, maize, and other traits found in the large, socially complex societies of the Americas. Writing in the 1960s, Ford saw a suite of innovations spreading (diffusing) throughout the vast area between the central United States and Peru, beginning about five thousand years ago. He subtitled his monograph "Diffusion, or the Psychic Unity of Mankind," asking readers whether they truly believed making pottery or building mounds was likely to be independently invented over and over again—hardwired, as it were, into the *Homo sapiens* brain. Laid out on maps, his data obviously could have been transported by travelers from region to region. Thirty years later, John Clark updated critical sections of Ford's monumental work. At first glance, it looks as if Ford was wrong: what in the 1960s had seemed to be a clustering of appearances of ceramics, mounds, and other features, now had many more early instances and a wider range of dates. Judging from the richer database, it seems unlikely that colonizers voyaged across the Caribbean in the third millennium B.C.E. carrying ceramics. Instead, Clark's study, with its maps peppered with dots, suggests multiple movements, of individuals as well as groups, introducing inventions or crops. Clark rejects the idea that there was a so-called mother culture back there in Olmec times, around 1500 B.C.E., from which historic cultures developed or budded off. His own research on the Olmec period in Mesoamerica indicates more than one political center, trading and borrowing ideas and maybe at times invading one another.

Joseph Needham insisted that the burden of proof lies more with claims of independent inventions of a trait than with hypotheses of intersocietal contacts, because we have an abundance of cases of innovations resulting from intersocietal contacts and only a few unequivocal cases of independent inventions. Kroeber documented a recurring pattern of *simultaneous* inventions, where necessary parts are already invented and there is strong interest to create a desired product. Railroads, steam engines, and personal computers are familiar examples of simultaneous inventions, often leading to competing claims and patent litigation. The material Ford and Clark amassed includes not only inventions but crops (maize and manioc) which, like the sweet potato, are organisms that were bred from wild plants growing naturally only in certain regions. Varieties cultivated outside the natural environment must have been carried—diffused—

by humans and, because most of the alien lands were already long inhabited, instituted through intersocietal contacts.

■ ■ ■ WAS OLMEC A SINGLE "MOTHER CULTURE" FOR MESOAMERICA?

Just when Clark's update of Ford's tables was published, an acrimonious debate erupted over Olmec trade. James Stoltman, a newly retired professor of North American archaeology, returned to his first love, laboratory analysis of ceramics, undertaking petrographic thin-section examinations of potsherds for colleagues. Stoltman would prepare paper-thin narrow slices cut from an excavated potsherd, mount them on glass slides, and use a microscope to identify the minerals in the thin-section; these would then be matched to geological occurrences of the minerals, usually pointing to clay sources utilized by the potters. Archaeologists interpret these data to indicate whether the pot in question was locally made or brought in from another area, indicating trade or perhaps intermarriage. The size of the sample (in this case of pottery) and precautions taken to avoid bias selecting the sample affect how valid the interpretation may be. In the case under dispute, Stoltman analyzed twenty sherds from Oaxaca, an area west of the homeland of the Olmec, and interpreted them as locally made in Oaxaca, although resembling styles common in the Olmec heartland site of San Lorenzo. Stoltman and his collaborators knew their interpretation was at odds with results from laboratory research employing a different technique, instrumental neutron activation analysis (INAA), carried out on a sample of nearly a thousand sherds from San Lorenzo and contemporary sites elsewhere in central Mexico. That study had indicated that *all* the *examined* Olmec-style pottery was manufactured from San Lorenzo area clays—not one piece made elsewhere, so far as the mineral analysis could show.

A much bigger question lurked behind the potsherds. "Olmec" is the name given by archaeologists to the earliest sites in tropical central Mexico beginning around 1200 B.C.E. that exhibit monumental architecture, art (Figure 7.1), agriculture, and other hallmarks of class-stratified complex societies—what most people label "civilization." Concepts first seen together in Olmec sites, including maize (corn) agriculture, stone statues of rulers and deities, pyramids, and icons of jaguar, feathered serpent, and its mouth as a cave, continued in most of the later Mesoamerican kingdoms. San Lorenzo archaeologists, and many colleagues, inferred from their data that San Lorenzo and nearby ancient towns played a central role in the development of Mesoamerican civilizations. Stoltman's analysis indicated that fine ceramics in the "Olmec style" (Figure 7.2) had been manufactured by another Mexican kingdom as well, implying that San Lorenzo Olmec had been only one among several early nations sharing innovations.

The heated debate illustrates how some trivial objects, in this case potsherds, become the crux of controversy over larger issues of cultural development. San Lorenzo archaeologists defended their interpretation by attacking the adequacy of

FIGURE 7.1 Monumental head of an Olmec king, Mexico, ca. 800 B.C.E. *Courtesy Milwaukee Public Museum.*

Stoltman's sample (his 20 versus their 944 sherds, plus their 123 samples of San Lorenzo area clays in the INAA study). Stoltman's sample was biased in that it comprised only foreign-looking sherds excavated in Oaxaca. The San Lorenzo study similarly selected Olmec-style sherds from collections recovered from sites outside the San Lorenzo zone, but it added an analysis of a good sample of San Lorenzo clays to see whether they matched the selected sherds. That they did seemed to clinch the interpretation that San Lorenzo manufactured and exported pottery but did not import much. Each side questioned how accurately the sherds were provenienced: how sure were the analysts that these pieces had been excavated from the sites claimed, and from San Lorenzo–period occupations, rather than from rodent burrows or from soil displaced by later building? The debaters attacked each other on scientific methods (INAA claims to register every mineral in a sample, and its practitioners assert that thin-section cuts can miss some inclusions), on whether samples were adequate, were selected without preconception bias, and were truly from careful, professional excavations.

Going beyond the technical arguments, the debaters openly fought over the role of San Lorenzo in Mesoamerican history. The San Lorenzo group was accused of espousing the old idea that civilization arose once, in a community of brilliant innovators whose striking accomplishments in art, architecture, and other technologies were widely imitated—San Lorenzo Olmecs as the "mother

FIGURE 7.2 Olmec-style pottery, Mexico, ca. 800 B.C.E. *Courtesy Milwaukee Public Museum.*

culture" of all subsequent Mesoamerican kingdoms. No, no, the San Lorenzo workers answered, we never said that, we just demonstrated objectively that this first great city exported a lot of pottery and imported little. The "Oaxaca group's" Early Olmec Period sites don't have San Lorenzo's city-scale settlement and architecture, or its colossal heads sculpted from hard stone transported from distant mountains, or its other large sculptures and fine art. We can't tell (yet?) whether trade, or aggression, related San Lorenzo to smaller comtemporaries across Mesoamerica. We can infer that from 1200 to about 850 B.C.E., every cosmopolitan person in Mesoamerica was aware of the kingdom with its impressive capital at San Lorenzo, and it seems reasonable that many desired an affordable product such as a bowl from the metropolis. Only further explorations for sites in the central Mexican tropics will illuminate relationships in the Olmec era when more complex societies emerged from a broad base of growing populations and whether technologies were diffused from a single source or were developed simultaneously in many adjacent areas.

■ ■ ■ BOATS

One of the questions in Olmec research is how the San Lorenzo people transported huge blocks of stone, weighing many tons, from mountain quarries to their lowland plateau. Did they use rafts, crossing open ocean to reach rivers flowing past their city? Could they maneuver heavily loaded vessels through the coast's mangrove swamps? Boats and water transport are involved in many archaeological controversies. Not many archaeologists are experienced sailors, and Western culture tends to see water travel as more dangerous than travel by land. Sailors, on the other hand, will maintain that open sea is less hazardous, though coastal waters, they do admit, are often perilous. Whatever your opinion, the fact is that Paleolithic immigrants to Australia floated themselves across miles of open ocean, and archaeology proves that at the end of the Pleistocene, people in Greece were going over to Mediterranean islands for obsidian and fishing, and Paleoindians over the Bering Strait.

Westerners' unfamiliarity with shipbuilding traditions other than that of Mediterranean Europe contributes to the controversies over probabilities of voyages. Cultural geographer Edwin Doran, Jr., recognized three great boat traditions, which he named *nao*, *junk*, and *vaka* (Figure 7.3). These derived from earlier boats that would have been log dugouts, rafts, bundles of reeds lashed together, or frames with a sewn covering of hides or bark. All these earlier types have been proven capable of crossing oceans (Table 7.1). *Nao* is a term derived from an Indo-European language root word for boat, *junk* is a familiar name for

Chinese ships, and *vaka* derives from variants of the Polynesian term for canoe. Few very ancient boats are directly known, because they were made of wood and thus have decayed, so controversies focus on comparing boats known ethnographically and on experiments with reconstructed vessels, evaluating their capability to carry people across large bodies of water.

Thor Heyerdahl, a Norwegian thoroughly experienced with all manner of boats, carried out the largest number of hands-on experiments with pre-Columbian sailing. He had an entrepreneurial spirit that, to his credit, he employed to fund reputable archaeologists for projects on Easter Island and in Peru, as well as to pay for his own voyages. Neither the Easter Island nor Peruvian excavations produced evidence for his pet hypothesis, that Polynesia was colonized by American Indians from South America, although archaeology did demonstrate that Peruvians sailed to the Galápagos Islands a few centuries before Spanish conquests. Heyerdahl's principal contribution was his experiments with reed boats and centerboard-equipped (a kind of rudder) balsa-wood rafts. Though such watercraft persist only on a small scale in a few areas today, their potential was demonstrated once Heyerdahl commissioned replicas big

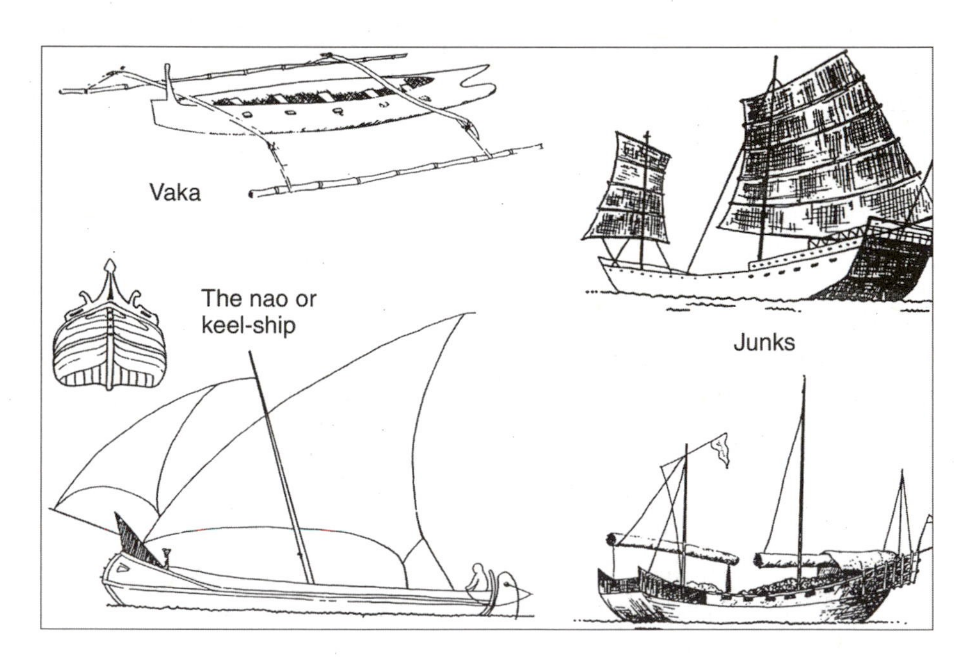

FIGURE 7.3 Three principal types of boats: *Top left*, Polynesian outrigger canoe; *bottom left*, European keeled *nao*; *right*, Chinese junks. *From E. Doran, 1973*, Nao, Junk, and Vaka: Boats and Culture History.

Table 7.1 Transoceanic Small Craft Crossings

Scientific Experiments	
1947	Thor Heyerdahl and crew on the *Kon-Tiki* balsa raft, Callao, Peru, to Raroia, Tuamoto Islands (eastern Polynesia), 101 days.
1970	Thor Heyerdahl and crew on the *Ra II*, reed-bundle boat, Safi, Morocco to Barbados (Caribbean), 57 days.
1976–77	Tim Severin and crew on the *Brendan*, Ireland to Newfoundland. Wintered in Iceland when three weeks of autumn storms made it too late to attempt sailing to Newfoundland.
Record-Breaking Small Boat Trans-Atlantic Attempts	
1866	The sloop *Alice* (48 feet), with owner T. C. Appleton, Captain Clarn, three seamen and a cook, Boston to Isle of Wight, 19 days.
1868	*Nonpareil*, three rubber cigar-shaped floats each 26 feet long, with light planking deck, two masts, carrying pump to reinflate floats, with John Mikes, George Miller, Jerry Mallene, New York to Southampton, 43 days.
1876	Danish-American Alfred Johnson rowed 26-foot dory, Gloucester, MA, to Abercastle, Wales, 57 days.
1891	J. W. Lawlor, alone, in *Sea Serpent,* 15 feet, with spritsail, Boston to Cornwall in 45 days.
1928	Captain Romer (captain on German-American Line ships), Cape St. Vincent, Portugal, to St. Thomas, Virgin Islands, on kayak made of waterproof fabric over wood frame, 90 days.
1952–53	Ann Davison, English, on 21-foot 6-inch sloop *Felicity Ann,* Canaries to Dominica, West Indies. Mrs. Davison and her husband had intended to cross the Atlantic on the *Reliance,* but both boat and husband were lost at sea a few miles out. She then sailed alone on the *Felicity Ann.*
1952	Dr. Alain Bombard, in 15-foot rubber dinghy *l'Hérétique,* living entirely off of caught fish, 65 days from Casablanca via Las Palmas to Barbados.
1980	Gerard d'Aboville, a Breton, rowed from Cape Cod to Ouessant, France, 3320 miles, in an 18-foot boat in 72 days, the first documented solo crossing from mainland to mainland.
1999	Tori Murden (36-year-old woman) rowed a 23-foot boat, *American Pearl,* 3000 miles from the Canary Islands westward to Fort-du-Bas, Guadeloupe, in 81 days; first woman to row alone across Atlantic. Murden was also the first American and first woman to ski to the South Pole.

Table 7.1 Transoceanic Small Craft Crossings (*continued*)

Trans-Pacific Crossings	
1882	American Bernard Gilboy on the 19-foot schooner *Pacific,* San Francisco to 40 miles NE of Sandy Cape, Australia (6500 miles) in 162 days. He lost sail and rudder, so a schooner brought him to land.
1972	John Fairfax and Sylvia Cook rowed 8000 miles in a 35-foot boat from San Francisco, drifting down the coast to Mexico before crossing to Hayman Island on the central Australian coast. Fairfax had rowed from the Canary Islands to Florida in 180 days in 1969.
1987	Ed Gillet paddled a kayak from Monterey Bay to Maui, Hawaii, in 63 days.
1991	Gerard d'Aboville rowed a 26-foot boat from Japan to Ilwaco, Washington, in 134 days.
1999	Kenichi Horie (60 years old), 103 days, 6800 miles from San Francisco to western Japan on sailboat made of 528 empty stainless steel beer kegs with sails made of recycled plastic bottles; in 1996, Horie crossed on solar-powered "yacht" made of melted-down aluminum beer cans; in 1962, 23-year-old Horie crossed on 19-foot yacht. All were solo crossings. *(Milwaukee Journal Sentinel* 7/9/99)
2001	Jim Shekhdar, 54-year-old British former computer salesman, rowed alone from Peru to Brisbane, Australia, 274 days (June 2000–March 30, 2001), 8060 miles, fastest Pacific crossing. (*Milwaukee Journal* 3/31/01)
Solo Around-the-World Trips	
1895–98	John Slocum, sea captain, out of Gloucester, MA, on the 36-foot 9-inch *Spray.*
1901–04	J. C. (Johann Klaus) Voss, naturalized Canadian, in the 50-foot red cedar Indian canoe *Tilikum,* three masts; for sections of the sailing, Voss had a second man on board (one swept overboard).
1942–43	Vito Dumas, of Argentina, on 32-foot ketch *Legh II,* 13 months 2 weeks (fastest circumnavigation by 1954).
Smallest Craft Used	
1981	Gerry Spiess made a 7800-mile Pacific crossing to Sydney in 5 months in a 10-foot sailboat; he had previously crossed the Atlantic in the boat.
1984	Arnaud de Rosnay disappeared at sea from a sailboard going from China to Taiwan. Earlier, his longest of seven open-water crossings was 1000 kilometers from the Marquesas to Ahé in the Tuamotus.

Table 7.1 Transoceanic Small Craft Crossings (*continued*)

1985	Two Frenchmen took 39 days to cross the Atlantic on a surfboard with a 20-inch-high hold for sleeping (one at a time).
1991	British sailor Tom McNally sailed from Portugal to San Juan, Puerto Rico, in a 5-foot 4-1/2-inch boat.
1993	Hugo Vihlen, American, beat that on the *Father's Day*, a 5-foot 4-inch boat, 106 days, St. John's, Newfoundland, to southern England.
2003	Raphaela Le Gouvello, French, windsurfed from Peru to Tahiti in 89 days.

enough to carry crews, supplies, and sails. His famous voyage from Peru to eastern Polynesia on the raft *Kon-Tiki*, in 1947, confirmed Spanish invaders' descriptions of indigenous merchants' sailing rafts from Ecuador to Chile. Heyerdahl next turned his attention to Egyptian boats, working from tomb paintings and models. No one in Egypt in Heyerdahl's day could build a reed boat, so he engaged Aymara Indian builders from Lake Titicaca in Bolivia, and in 1970 crossed the mid-Atlantic in *Ra II*, taking fifty-seven days. *Ra* became waterlogged down to her gunwales, and this made her so heavy she dragged through the ocean, but *she did not sink*. After the voyage, Heyerdahl visited Marsh Arabs in Iraq, who use reed boats and told him that he should have cut his reeds in August, not in winter, and they would have remained buoyant much longer. *Kon-Tiki* and *Ra* revealed that experienced boatbuilders who kept their supplies higher than the deck could cross oceans on rafts or reed boats.

Meanwhile, another sailor, Tim Severin, with National Geographic Society support tested the feasibility of ocean crossings on hide-covered wooden-frame boats. Irish legend says that St. Brendan made a long voyage in such a boat, at least to Iceland and maybe to America. Scoffers laughed that the hides would leak at the seams or get waterlogged quickly. Severin studied Roman descriptions of the big merchant *curraghs* traversing the Irish Sea with cattle and grain for Roman English markets. Today's small curraghs, in western Ireland, and coracles in Wales, use heavy tarred canvas instead of hides. Severin did succeed in sailing his Roman-era-sized hide-covered boat, the *Brendan*, across the North Atlantic. He and his crew pulled up on each of the North Atlantic's main islands to grease *Brendan*'s skins (a stinky job) and caulk seams—no doubt St. Brendan's own practice—and the modern crew wintered in Iceland to avoid icy storms. Even though the *Brendan* was designed watertight, unlike the wash-through rafts and reed boats, she proved remarkably stable.

Heyerdahl and Severin put sails on their ships, referring to ancient models and pictures. The earliest direct evidence for sails is from five thousand years ago, and their advantage is considerable, producing greater speed with little human energy input. In fact, the sail is probably the first, and for a long time the greatest, invention for harnessing mechanical energy. The force driving a sailing ship through waves can be tremendous. Europeans strengthened plank-built ships with a heavy timber keel lengthwise under the middle of the ship, plus stern and stem posts and bracing ribs. Chinese boatbuilders did something quite different, constructing a broad, flat plank bottom, strengthening the ship with strong timber gunwales, and making bulkheads instead of using ribs. European ships are pointed at the ends; Chinese junks are squared off. European *nao*-type ships look streamlined to us, but junks seem to go through water just as well, and their wide, flat bottoms give them greater stability and more cargo capacity. Oceanic *vaka* canoes can be one hundred feet long, with stability created by double hulls or outriggers to give broader beam.

Nautical technology has a very long history, which in itself increases the probability of significant human movements across water (with plenty of centuries in which to experiment and make refinements). The technology of shipbuilding was brought together with information technology to understand winds and currents. As Heyerdahl reiterated, when sailing over water, mileage on a globe or map doesn't mean much—it's the direction and speed of winds and currents that make sea journeys long or shorter. Traveling over water didn't leave trails or traces the way land travel did. Archaeologists could only investigate where journeys might begin and end, not voyages themselves. Nautical archaeology, which works on offshore wrecks, not boats or items lost in deep ocean, began only in the 1960s, and is expensive to mount and very time-consuming. Frustrated from studying marine travel directly, archaeologists seemed to project upon ancient people the notion that water is an insurmountable barrier, though ancient sailors may not have shared this fear.

■ ■ ■ PRE-COLUMBIAN TRANS-PACIFIC CONTACTS

Polynesians on every habitable island in the eastern Pacific centuries before Columbus irrefutably prove that Asians could, and would, cross the ocean. A few scholars have pursued similarities between Asian and American cultures, accepting the demonstrated capabilities of junks, Polynesian ships, and rafts to travel the Pacific. As mentioned above, the sweet potato is a link that cannot be denied. Geographer Carl Johannessen has compiled a list of more than two hundred organisms, both plant and animal (including insects and disease pathogens), evidencing pre-Columbian transoceanic contacts between the Americas and Eurasia or Africa. Most archaeologists don't want to think about this evidence; it makes them uncomfortable to discuss data drawn from research areas outside their own studies, and it's hard to overturn the reigning paradigm that "primitive people" couldn't cross water "barriers." Complicating the situation for scholars seriously interested in examining this question is the influence of avocational enthusiasts who believe that sunken continent refugees, from Atlantis or Mu, jump-started American civilizations. This group of enthusiastic non-professionals has no problem positing transoceanic contacts from the vanished golden cities, with or without concrete evidence. Students wanting to study transoceanic diffusion are told by their professors to simply avoid dealing with any proponents of pre-Columbian ocean crossings.

Against the mainstream, your author here has seriously considered transoceanic and transcontinental contact evidence since she was a student aide assisting James Ford and Gordon Ekholm, first-rank archaeologists advancing carefully researched hypotheses on intersocietal contact over major bodies of water. I think the evidence for a number of cases is compelling. Complexes that agree well with Needham's principle of collocation include the following:

- Paper-making. In both China and Mesoamerica, bark of the deliberately planted paper mulberry tree was soaked, beaten with stone mauls, and dried in sheets to make paper, and not only used for writing but also cut into shapes, like paper dolls, to represent spirits in funerals and other rituals. On islands of the Pacific between China and Mexico, the process was carried out to the point of producing bark cloth but stopped short of the additional steps to make paper. Stone mauls discovered in archaeological sites in South America and the Pacific islands look like those used historically for bark beating.

- Royal purple dye. It is made from a cream-colored secretion from a marine snail, which turns purple during the manufacturing process. Other dyes, such as scarlet from dried cochineal insects on cactus, blue from indigo leaves, and red from madder root, might have been independently discovered, but the multi-step process of making "Tyrian purple" from the whitish secretion from a marine snail suggests borrowing. The other brilliant dyes mentioned may also have been transmitted by transoceanic contacts.

- Wheeled figurines. Found in Mesoamerican archaeological sites, clay animals with axles through their feet ending in wheels have been dismissed

> by American archaeologists as toys, although many were found in adults' tombs. Indigenous Americans did not use wheels either for transport or for throwing pots in making pottery. Wheeled animal figurines were commonly placed in Chinese tombs to represent sacrifices. Since wheeled figurines and paper are both associated with funerals, and have long histories in China, it is reasonable to hypothesize that both manufactures were transmitted, possibly together, from Asia to Mexico.

Dozens of other parallels can be listed. Some of them, such as breeding fighting cocks and putting Asian-style spurs on them, or distilling liquor through Asian-type pot stills, may possibly have been transmitted very soon after the Spanish conquest of Mexico by Asian seamen deserting from horrendous conditions on the trans-Pacific Manila galleons sailing regularly between Acapulco and Manila. Any seaman fleeing his ship before completing a round trip was severely punished, so he would disappear into the hinterland as fast as he could, to live with a native community beyond Spanish control.

Other links are definitely pre-Columbian. Cocaine has been identified in Egyptian mummies of the first millennium B.C.E. by a forensic chemist: cocaine has to come from the South American coca plant, and its residue in a body is a complex molecule that in a court of law would send a dealer to prison. A number of customs in Mesoamerica are particularly Asian-like, including valuing jade highly and placing a jade bead painted red with cinnabar in the mouth of corpses; building tiered pyramids symbolizing the seven or nine or thirteen heavens; formal body and hand positions (called *mudra* in India) seen in Mesoamerican art, and the ruler seated on his throne with one leg tucked under, the other hanging. Other customs, like calendar astrology (see below), are widespread in Eurasia, including the Tree of Life with a great bird on top, a lion or jaguar on earth, and a serpent among its roots. Still others are more contested, including the identification of ears of maize (corn) carved on medieval temples in South India, and peanuts—a South American crop—recovered from archaeological excavations in China. Peanuts and other American crops continue to turn up in Eurasian and Oceanic excavations, substantiating references in pre-Columbian Chinese and Indian texts on cultivated plants. Taken together, there is strong evidence for multiple transoceanic contacts and borrowings before 1492, and at the same time, absolute evidence that American civilizations developed independently of any attempts at colonization by Eurasian or African nations. What the evidence shows is that America's indigenous nations were part of global connections for several thousand years before Columbus kicked off the historic invasions.

Calendar Astrology

An obstacle to "thinking out of the box" is that it may require an effort to understand unfamiliar technical matters. A case in point is what, in my opinion, is perhaps the most compelling evidence for pre-Columbian trans-Pacific contacts: systems of calendars linked with astrological predictions. Alexander von Humboldt (1769–1859), a pioneering naturalist, explored throughout Latin America early in the nineteenth century. He published his observations on indigenous peoples and archaeology in 1814, describing at length many similarities between native American cultures and those of Asia. A majority of scholars pooh-poohed the possibility of transoceanic borrowings, insisting the oceans are too wide to have been crossed before fifteenth-century European ships were available. Ignoring Polynesian voyagers, the legion of doubters have disregarded von Humboldt's carefully detailed comparisons, disrespecting a researcher acclaimed as one of the most brilliant scientists of his century. Among von Humboldt's discoveries are astonishing parallels between Mesoamerica and Asia in astronomy, astrology, calendar systems, and cosmology.

The easiest to describe among these parallels is the image that Mesoamericans and Chinese saw in the full moon: a rabbit pounding something in a mortar! The Chinese said he is pounding plants to make the elixir of immortality; the Aztecs said he is pounding maguey to make *pulque*, their beer. Take a look at the next full moon. Do you see a rabbit on its hind legs, pounding with a pestle into a mortar? If you do manage to make out such a figure, is it so obvious that societies on the two sides of the Pacific should both see it? Is it somehow so obvious, yet no other societies have seen that rabbit and mortar?

More fantastic than the rabbit drinkmaster in the moon are the calendar system and associated astrology parallels between Asia and Mesoamerica. We can begin with projecting into the night sky a set of constellations, each in its "house," one for each day of the twenty-eight that the moon is visible every month. These are the "lunar mansions" of astrology, the moon spending a day in each "house" as it moves through the sky during its monthly phases. Then we have a zodiac of twelve constellations along the ecliptic, the zone in the sky in which eclipses take place. The apparent path of the sun and the orbits of the visible planets lie in this zone. Astrologers inquire the exact moment of a person's birth and calculate which stars and planets were ascendant at that moment, ascribing characters to the stars and planets that supposedly will be part of the person's character (you can read these horoscopes carried in most newspapers even today).

Astrology developed in Mesopotamia at least by the first millennium B.C.E., patronized by royalty (also by the late President Ronald Reagan and his wife Nancy) to divine future threats and plot courses of action to nullify them. Why would anyone premise that a person's character and fate would be associated with stars and planets in the ascendant during the moment of birth? Learning to read directions and time at night through the movements of stars and planets could be independently invented, by sailors, herders, war parties, or travelers in the desert, but there doesn't seem anything inherent in the night sky—much less stars invisible during the day—to suggest a key to a person's life. Yet Mesoamericans not only divided the sky into lunar mansions and zodiac, they also cast horoscopes, and saw in the night sky iconic constellations parallel to many Eurasian ones (for example, Capricorn = "goat-fish" or swordfish in Eurasia, swordfish or crocodile in Mesoamerica).

Mesoamerican zodiacs and lunar mansions don't have all the same animals or deities as those in Eurasia, which in any case differ from one from another, too. After all, Eurasian bulls, lions, goats, and horses didn't live in the Americas. Mesoamericans and western Eurasians did both recognize four basic elements: earth, air, fire, and water (the Chinese added metal to make five). Further strong parallels exist between Mesoamerican and Asian named days in the calendar, associated with colors, elements, and characters. The seven days of our week are named for seven visible planets, Saturn being the last (Saturday) and the sun (Sunday) treated as a planet. The order of the planets (including the sun) is not clear to ordinary observers, and they certainly aren't likely to postulate an invisible eighth and ninth planet that is a double-headed dragon. On the coffin in the tomb of Lord Pacal (A.D. 603–683) of the Mayan kingdom of Palenque (see Figure 3.1 in Chapter 3) is a series of hieroglyphs reading Sun, Moon, Mars, Mercury, Jupiter, Venus, Saturn—*the same order as our Eurasian week*: Sunday, Moonday, Zeus [Tues]day, Wodensday [Anglo-Saxon Woden = Roman Mercury], Thorsday [Anglo-Saxon Thor = Roman Jupiter], Frigsday [Anglo-Saxon Frig = Roman Venus]. Nine-day weeks incorporating the double-headed dragon as the last two days are known from India, China, and Mesoamerica, as well as the more general seven-day week.

Mesoamericans had twenty-day months (easier to calculate with than our months of differing numbers of days), so there isn't an exact parallel to Asian twenty-eight-day months, but there are *sequences* of day names and associations suggesting borrowing (Table 7.2). Among the striking and unexpected parallels is the Aztec day name "broom plant" which corresponds to the Chinese lunar mansion "willow" (willow shoots were made into brooms). China and Mesoamerica shared the complication of two simultaneous calendars, of differing lengths, that meshed like cogwheels, arriving at the same day starting point every so many years—52 for Mesoamerica, 60 for China. Westerners today are familiar with the Chinese custom of naming years after zodiac animals, in a twelve-animal sequence that repeats. Zodiac animals are also used, more or less in standard sequence, as names of lunar mansions. They appear, significantly

Table 7.2 Aztec Day Names Compared with Chinese and Greek Lunar Animals

Aztec	Chinese	Greek
1. Crocodile or Swordfish	1. Dragon	9. Snake or Dragon
2. Wind		
3. House	Purification Temple Constellation	
4. Lizard	4. Serpent	9. Snake or Dragon
5. Snake	4. Serpent	9. Snake or Dragon
6. Death	Piled-up Corpses Constellation	
7. Deer	5, Little Deer; 7, Big Deer	20. Deer
8. Rabbit	27. Hare	
9. Water	Gemini Constellation "Accumulated Water"	
10. Dog	9. Wild Dog; 15, Dog	7. Dog
11. Monkey	10. Little Monkey; 11. Big Monkey	15. Baboon
12. Twisted [broom plant]	Lunar Mansion *liu* "Willow"	
13. Reed		
14. Ocelot [a feline]	24, Leopard; 25, Tiger	16, Cat; 17, Lion; 18, Leopard
15. Eagle	12. Crow; 13, Cock; 19, Swallow	5. Hawk
16. Vulture	12. Crow; 13, Cock; 19, Swallow	2. Vulture
17. Earthquake		
18. Flint		
19. Rain, or Turtle	Tortoise Constellation	
20. Flower		

Source: Kelley 1960.
Note that there cannot be exactly matching lists because Mesoamericans had twenty-day sequences and Eurasians had twenty-eight-item (lunar animals) sequences.

omitting Eurasian domesticated animals, mostly in standard sequence as Mesoamerican day names in their twenty-day months. Since naming calendar days after animals and putting these in a particular sequence seems to have no evident basis in natural phenomena—even the constellations of the zodiac and lunar mansions are largely arbitrary configurations—to find sequences of animal names for days, with associated characters, colors, and elements, plus the same seven-day sequence of planets, on both sides of the Pacific beginning about two thousand years ago, strongly implies contact. All the calendars depend upon close long-term observation of the skies and some way of keeping notes and performing calculations, yet the scientific astronomy involved need not have given rise to concepts of animals in lunar mansions and zodiacs, sequences of day names with associated color, element, and character, and astrological imputations for human fates. David H. Kelley, who has devoted decades to pondering, calculating, and comparing Mesoamerican and Eurasian calendar astronomy and astrology systems, considers the link between the two to have occurred about two thousand years ago, between northern India and Mesoamerica, probably Guatemala; in his view, one educated Mesoamerican learned the complicated principles and mythological associations of Eurasian systems and reworked them into his own knowledge, improving upon the Eurasian calendars with a more sophisticated manner of handling the discrepancy between fixed calendars and true solar years than Eurasian leap years. Did contact also bring over shared concepts of sequential destructions of the world, a deity sacrificing itself by jumping into a fire to rejuvenate humankind, and the remarkable parallel between three sequential Mayan day names *Ceh-Lamat-Muluc* and Hebrew alphabet *kaph-lamed-mem*/Greek *kappa-lambda-mu*, with the first represented by a hand by both Hebrew and Maya, and the third associated with water?

Norse in America, 1362

Far to the north of Mesoamerica, evidence for quite another kind of transoceanic contact was discovered a century ago by a Swedish immigrant to northwestern Minnesota named Olof Ohman. Late in August 1898, Ohman was grubbing up trees on his homestead farm, using a Swedish technique of chopping the lateral roots of trees and winching the tree out of the ground. Ohman's ten-year-old son carried some lunch out to him and stayed to help. A large flat stone came up wedged between the roots of one of the trees as its stump was pulled out of the ground. The boy noticed odd lines on the stone, and rubbed the dirt off. His father had had very little schooling back in Sweden, but he recognized the writing as Swedish runes, and called over his neighbor who was grubbing trees on his adjacent land. Ohman brought out a wagon, and they carried the stone and the stump to his farmyard. All the neighbors around their village of Kensington, nearly all immigrants from Scandinavia, came over during the next few days to see the stone with the rune inscription, and the stump with its roots bent to go around the stone (Figure 7.4).

One of the Kensington residents copied the inscription and sent the copy to the University of Minnesota, where it was forwarded to a professor of Scandinavian languages. This man detected what he considered errors in the language and shapes of some of the runes, and declared it a fraud. Other professors agreed. The matter was dropped in the face of these experts' disapproval. Several years later, a young man working his way through graduate study in history at the University of Wisconsin came by. He was Norwegian-American and spent summers selling maps and books to Scandinavian homesteaders in the Old Northwest (Wisconsin and Minnesota). At the University of Wisconsin, he had studied Old Norse language and runes, and when told of Ohman's find, he went out to the farm, read the stone, and was convinced it probably was authentic. It read:

> Eight Götlanders and twenty-two Norwegians on an acquiring journey to the west from Vinland. We had camp by two shelters one day's journey from this stone. We were fishing one day. After we came home, we found ten men red

FIGURE 7.4 The Kensington Runestone, with rune inscription carved by Norse explorers in Minnesota, 1362. Olof Ohman, the farmer who found it, at right beside it, 1927. *Courtesy Ohman family.*

> from blood and death. Ave Maria save from evil. There are ten men by the sea to look after our ships fourteen days journey from this island. Year 1362. (translation by Richard Nielsen)

The young historian commenced what would become a lifelong campaign to change the judgment of fraud on the stone. His efforts were in vain; he died in 1963, author of several books giving positive arguments for the so-called Kensington Runestone, yet he was mocked as a fanatic. Twenty years later, a Danish-American engineer working in Scandinavia took a copy of the Kensington inscription to a professor of medieval Norse in Copenhagen. This scholar read the text, looked up and said, "This is fourteenth-century Bohuslän [Swedish] dialect. Where did you get it?" His visitor, Richard Nielsen, replied, "Minnesota."

Back in the United States, Nielsen and another engineer, who found him by looking on the Internet, agreed the Runestone should be examined in a modern high-tech petrographic laboratory. Hard data by a competent geologist should settle the question of whether weathering indicates the runes were carved in the nineteenth or the fourteenth century. In 2000, this examination was performed by a forensic petrographer (a geologist called in as an expert witness in legal cases). The evidence was straightforward: the edges of the rune carvings are too weathered to have been recently carved when Ohman produced the stone for his neighbors' wonderment. The inscription, by the weathering evidence, must have been at least 200 years old in 1898. Euroamerican settlement began in northwest Minnesota only in the 1870s, after the Civil War. The petrographer's data could not prove the inscription was made in 1362, but if we reject the nineteenth-century hoax explanation, the sensible next step is to entertain the hypothesis that the inscription is genuine, carved by a party of Norsemen in 1362, as it says. This is the most reasonable way ("Ockham's Razor"; see Chapter 2) to account for its presence.

Nielsen had asked me to advise him on any archaeological aspects of his investigation. There wasn't much of that, as archaeological tests of the knoll from which the stone had come (the knoll is almost an island rising from marshy surroundings) had produced only flakes from stone tools, indicating probable brief camps by Indians. Of course, the inscription doesn't say the Norse camped on the knoll. If they did, expecting to uncover Norse metal tools or ornaments in narrow trenches widely spaced over the knoll would be like expecting to find a needle in a haystack, *if* the needle had not been carried away as too valuable to lose. I did advise Nielsen that Scandinavia in the early 1360s, not long after the terrible Black Death plague had killed nearly half the people, was turbulent, and that Germans' hostile takeover of the Norse countries' profitable fur trade would be a reason for a party of Norsemen to travel far inland from Vinland on the Atlantic coast to acquire furs to sell. History also suggests that if any of the party returned home, he would have joined his nation's military campaign to oust the Germans, which succeeded in 1368. There is a rational historical reason for a journey westward into America in 1362, and a rational reason it seems not to have led to a continuing inland fur trade such as was initiated in 1670 by Britain's well-capitalized Hudson's Bay Company.

The majority of archaeologists and historians insist they know the Kensington Runestone inscription is a hoax. How can they "know" this when Olof Ohman and his neighbors swore affidavits on the truth of his unexpected find, and no one has been fingered who could have carved this extensive text on a very hard stone in barely colonized northwestern Minnesota? When the "errors" in rune shape and language turn out to be Swedish dialect variations not known in the nineteenth century, but found by twentieth-century scholars? Nay-sayers do not consider the extensive collection of papers in the Minnesota Historical Society including the affidavits, later letters from Kensington residents, and the conclusion of an eminent geologist in 1910 that, as was seen ninety years later, weathering on the runes indicates carving too long ago to have been done by post–Civil War settlers. One didn't need an electron scanning microscope to see that, although it is good that we have these higher-resolution data now from that instrument.

Could Norse have physically managed to get from the Atlantic, where they had been settled in southern Greenland since 982, inland to Minnesota? Norse men were accustomed to very long journeys through Russian forests and marshes to obtain furs to trade in Constantinople (now Istanbul, Turkey) and in southern Europe. We know, from carefully excavated settlement remains, that Norse had briefly lived, at A.D. 1000, in L'Anse aux Meadows at the northern tip of Newfoundland, Canada. For five hundred years, from 1000 to nearly 1500, Norse farmers in southern Greenland sailed over to Labrador in eastern Canada to obtain timber for buildings and to trade with Indians and Inuit there for furs and walrus hides and tusk ivory to sell to Norwegian merchants who voyaged to Greenland to sell European goods. Some Greenland Norse seem to have themselves hunted in arctic Canada, rather than depend on Inuit to meet them at the coast. In 1362, after stopping over in Greenland and talking with their countrymen there, a party of Norse could have sailed by boat up the St. Lawrence River and the Great Lakes to the area around Duluth, Minnesota, and then canoed fourteen days westward to the Kensington region. More likely (according to a paleo-geographer), they could have sailed into Hudson Bay and then up the Nelson or Hays River into Lake Winnipeg, and from Winnipeg, down the Red River into Minnesota, as the later Hudson's Bay Company traders did. Like historic fur traders, they would have relied on Indian guides familiar with First Nations' long-standing, continent-wide trade routes. With Nielsen's research into fourteenth-century manuscript runes and Old Swedish, and the petrographer's scientific hard data (literally), the *weight of probability* falls toward the Kensington inscription being a genuine record of a medieval Norse incursion seeking American furs.

Insistence that the Kensington rune inscription "must be" a modern hoax because medieval Norse "could not" explore North America fits a mind-set that considers extensive travel possible only with the advent of our complex, industrial or electronic technology. Until archaeological excavations at L'Anse aux Meadows in the 1960s exposed unmistakable foundations of medieval Norse houses, a boatshed, and iron-working, and a few Norse artifacts, in Canada, three sagas chronicling Greenland Norse efforts to colonize in America had been held to be myths. Once the archaeological proof of at least one attempt at settlement had been accepted, archaeological evidence for occasional Norse in arctic Canada was also accepted, although whether Norse, or only the Inuit they traded with, were at particular sites may be debated. We have abundant historical evidence for Norse fur-trading journeys through Russian forests similar to forests in northern Minnesota. Thus the argument that Norse in 1362 "could not" have reached Minnesota fails to recognize that both Scandinavian and Greenland Norse had technology and experience in sailing, traveling through northern forests, and collecting furs from native people. As far as technology goes, medieval Norse were as well prepared to explore into Minnesota as were the Hudson's Bay Company men three centuries later; the difference between the traders lay in capital, not technology: the English company had financial resources to carry on explorations for years before it turned a profit.

Evaluating Diffusion Claims

How do we distinguish between less viable and more viable claims? I don't want to sound elitist, but it does make a difference if the person earned an advanced degree from a reputable university *in the field of research under discussion*. Joseph Needham was called one of the greatest scientists of the twentieth century, elected Master of one of Cambridge University's colleges, and—more to the point, because his expertise had been focused on biochemistry—enlisted a panoply of highly respected Western and Asian scholars to collaborate with him and Dr. Lu on the volumes of *Science and Civilisation in China*. David Kelley, who researches Eurasian and Mesoamerican calendar astrology, wrote his accepted Harvard Ph.D. dissertation about trans-Pacific contacts, and his professional research in

Maya archaeology and epigraphy has been well received by his colleagues. These scholars, like James Ford, Gordon Ekholm, and Carl Johannessen, built reputations for sound scientific work through conventional research in addition to their less-accepted proposals. On the other hand, I have for many years worked with, listened to, and read papers written by avocational researchers, and observed that many good citizens endeavor to learn about a subject that interests them and to research it, but without professional guidance on methods and sources, it is easy to be misled by exciting but unscientific ideas.

It usually happens that researchers with good scholarly credentials and mainstream work find that their "thinking out of the box" is politely ignored. This is largely a result of the culture of the university. One colleague with whom I have discussed the problem remarked that people "learn stories" (such as Columbus discovered America) as they are growing up, and are uncomfortable when somebody "tells the story wrong." That seems the explanation for archaeologists properly trained in the scientific method who avoid talking about contradictions to what they learned as "facts" from their professors years ago. There is also legitimate worry that unorthodox opinions may jeopardize getting a job; the older generation who hires younger researchers and teachers may not welcome challengers to their ideas and procedures. Reluctance to go beyond one's primary area of research, a basic risk in examining evidence for long-distance contacts, is another caution inhibiting archaeological study of such data. Bringing in scientists with expertise in DNA, petrography, chemistry, astronomy, and similar "hard sciences" helps overcome archaeologists' reluctance to trust a colleague's argument. Sometimes, reluctance arises when a case involves data not usually part of archaeological study, nor suited to physical science techniques. To sum up here, the majority of archaeologists hesitate to evaluate arguments for transoceanic contacts earlier than 1492 because the sets of data include areas outside one's usual research, most people assume oceanic voyaging in small boats would have been hazardous, the Columbus-discovered-America story is familiar from childhood and seems reasonable, and challenging received opinion can be dangerous to one's financial health. Still, slowly the data for transoceanic contacts before 1492 mount up and, I predict, will eventually result in a paradigm shift.

SOURCES

Boehm, David A., Stephen Topping, and Cyd Smith, editors, 1983. *Guinness Book of World Records*. New York: Sterling.

Carlson, John B., 1984. The Nature of Mesoamerican Astronomy: A Look at the Native Texts. In *Archaeoastronomy and the Roots of Science*, edited by E. C. Krupp, pp. 211–252. Boulder, CO: Westview Press.

Clark, John E., and Michelle Knoll, 2005. The American Formative Revisited. In *Gulf Coast Archaeology: The Southeastern United States and Mexico*, edited by Nancy Marie White, pp. 281–303. Gainesville: University Press of Florida.

Doran, Edwin, Jr., 1973. *Nao, Junk, and Vaka: Boats and Culture History*. College Station: Texas A & M University.

Fell, Barry, 1976. *America B.C.* New York: Pocket Books.

Fingerhut, Eugene R., 1994. *Explorers of Pre-Columbian America? The Diffusionist-Inventionist Controversy*. Claremont, CA: Regina.

Ford, James A., 1969. *A Comparison of Formative Cultures in the Americas*. Smithsonian Contributions to Anthropology 11. Washington, DC: Smithsonian Institution Press.

Green, Roger C., 1998. Rapanui Origins Prior to European Contact: The View from Eastern Polynesia. In *Easter Island and East Polynesian Prehistory*, edited by P. Vargas Casanova, pp. 87–110. Santiago: Universidad de Chile, Instituto de Estudios Isla de Pascua.

Hägerstrand, Torsten, 1967. *Innovation Diffusion as a Spatial Process*. Chicago: University of Chicago Press.

Heyerdahl, Thor, 1978. *Early Man and the Ocean*. New York: Vintage (Random House).

Hodgen, Margaret T., 1974. *Anthropology, History, and Cultural Change*. Tucson: University of Arizona Press.

Jett, Stephen C. 2002. Nicotine and Cocaine in Egyptian Mummies and THC in Peruvian Mummies: A Review of the Evidence and of Scholarly Reaction. *Pre-Columbiana* 2(4):297–313.

—. Ancient Ocean Crossings: The Case for Pre-Columbian Contacts Reconsidered. MS. in the author's possession.

Jones, Terry L., and Kathryn A. Klar, 2005. Diffusionism Reconsidered: Linguistic and Archaeological Evidence for Prehistoric Polynesian Contact with Southern California. *American Antiquity* 70(3):457–484.

Kehoe, Alice Beck, 2003. The Fringe of American Archaeology: Trans-oceanic and Transcontinental Contacts in Prehistoric America. *Journal of Scientific Exploration* 17(1):19–36.

—. 2005 *The Kensington Runestone: Approaching a Research Question Holistically*. Long Grove, IL: Waveland Press.

Kelley, David H., 1960. Calendar Animals and Deities. *Southwestern Journal of Anthropology* 16(3):317–337.

—. 1981. The Invention of the Mesoamerican Calendar. Unpublished paper in possession of author.

Kelley, David H., and Eugene F. Milone, 2005. *Exploring Ancient Skies: An Encyclopedic Survey of Archaeoastronomy*. New York: Springer.

Kirch, Patrick Vinton, and Roger C. Green, 2001. *Hawaiki, Ancestral Polynesia*. Cambridge: University of Cambridge Press.

Kristiansen, Kristian, and Thomas B. Larsson, 2005. *The Rise of Bronze Age Society: Travels, Transmissions and Transformations*. Cambridge: Cambridge University Press.

Kroeber, Alfred L., 1952. *The Nature of Culture*. Chicago: University of Chicago Press.

Latin American Antiquity 17(1), 2006. San Lorenzo researchers (Hector Neff, Jeffrey Blomster, Michael Glascock, Ronald Bishop, James Blackman, Michael Coe, George Cowgill, Richard Diehl, Stephen Houston, Arthur Joyce, Carl Lipo, Barbara Stark, and Marcus Winter): 54–76; Oaxaca researchers (Robert Sharer, Andrew Balkansky, James Burton, Gary Feinman, Kent Flannery, David Grove, Joyce Marcus, Robert Moyle, Douglas Price, Elsa Redmond, Robert Reyolds, Prudence Rice, Charles Spencer, James Stoltman, and Jason Yaeger): 90–103; San Lorenzo researchers (Hector Neff, Jeffrey Blomster, Michael Glascock, Ronald Bishop, James Blackman, Michael Coe, George Cowgill, Stephen Houston, Arthur Joyce, Carl Lipo, Ann Cyphers, and Marcus Winter):104–118.

Merrien, Jean [real name, René Marie de la Poix de Fréminville], 1954. *Lonely Voyagers [Les Navigateurs Solitaires]*. English translation 1954, by J. H. Watkins. New York: G. P. Putnam's Sons.

Needham, Joseph, 1986. *Science and Civilisation in China*, vol. 5, pt. 7 (Military Technology; The Gunpowder Epic). Cambridge: Cambridge University Press.

Needham, Joseph, and Gwei-Djen Lu, 1985. *Trans-Pacific Echoes and Resonances; Listening Once Again*. Singapore: World Scientific.

Nielsen, Richard, and Scott F. Wolter, 2006. *The Kensington Runestone: Compelling New Evidence*. St. Paul, MN: Lake Superior Agate Publishing.

Nichols, Johanna, 1992. *Linguistic Diversity in Space and Time*. Chicago: University of Chicago Press.

Rogers, Everett M., 1962. *Diffusion of Innovations*. New York: Free Press.

Shao, Paul, 1976. *Asiatic Influences in Pre-Columbian American Art*. Ames: Iowa State University Press.

Severin, Tim, 1978. *The Brendan Voyage*. New York: Avon.

Sorenson, John L., and Carl L. Johannessen, 2006. Biological Evidence for Pre-Columbian Transoceanic Voyages. In *Contact and Exchange in the Ancient World*, edited by Victor H. Mair, pp. 238–297. Honolulu: University of Hawai'i Press.

—. 2004. *Scientific Evidence for Pre-Columbian Transoceanic Voyages to and from America*. Sino Platonic Papers 133, on CD-ROM only. Philadelphia: University of Pennsylvania, Department of Asian and Middle Eastern Studies.

Storey, Alice A., Jose Miguel Ramirez, Daniel Quiroz, David V. Burley, David J. Addison, Richard Walter, Atholl J. Anderson, Terry L. Hunt, J. Stephen Athens, Leon Huynens, and Elizabeth Matisoo-Smith, 2007. Radiocarbon and DNA Evidence for Pre-Columbian Introduction of Polynesian Chickens to Chile. *Proceedings of the National Academy of Sciences* 104(25):10335–10339.

Webb, S. [Steve] C., 2006. *The First Boat People*. [Pleistocene Australians]. Cambridge: Cambridge University Press.

8 What People Before Us Could Do
Earlier Technologies

Money-grubbing fantasists like "psychic archaeologist" Erich von Däniken (Chapter 3) proclaim our forebears couldn't build anything impressive, therefore famous ruins must be the legacy of civilizers from outer space. Until we have more than psychics' soul journeys to convince us, we must play it straight: we see the ancient constructions and manufactures, we know earthbound humans lived at these places, we as scientists hypothesize that humans made what we see ("Ockham's Razor," Chapter 2). Television documentaries show hearty, sweating guys experimenting with lifting big stones near the pyramids at Giza, or standing multi-ton obelisks upright, or launching galley ships with dozens of grunting rowers, while most archaeologists concern themselves with recording what exists and comparing those data to other sites. The basic fact is that human strength and ingenuity have produced a huge array of artifacts, from crude Paleolithic hammerstones to the Coliseum and St. Peter's in Rome. Controversies among archaeologists focus on interpreting *why* as well as how people made something, and whether the technology was introduced from another society or independently invented. Although we may not have the answer to "how" in each case, we have enough evidence that we needn't resort to fantastic theories about galactic astronauts or lost civilizations of giants.

Archaeoastronomy

Mathematics is a broad field encompassing geometry, algebra, calculus, statistics, and more. It was practiced in many ways by many peoples in the ancient world. Two thousand years ago, the Hopewell people in eastern North America, like their near contemporaries the Classical Greeks, were fascinated by geometry. We think of mathematical calculations as numbers and symbols progressing along a page or blackboard or computer monitor. People without notebooks did geometry by drawing lines and angles, or laying lines out on the ground. They could relate star movements (as they appear to us, but actually an effect of our earth's rotation) by observing geometric relations in the sky, and drawing or laying them out. Two thousand years ago, learned Maya in Mesoamerica were calculating star movements into a complex calendar that meshed a 260-day ritual "year" with a sun-based 365-day year (the 260-day period may have originally been the portion of the year the southern Maya did not need to work in their maize fields). Literate in their hieroglyphic writing system, the Maya tied calendar dates to historic events and back to mythic beginnings which they figured to be more than a million years earlier.

Maya astronomers calculated eclipses and the periods when certain stars would not be visible. They wrote tables of their calculations in reference books, of which only one, the "Dresden Codex," escaped burning by invading sixteenth-century Spaniards who believed the glyphs drawn as strange personages were pictures of the Devil. This one surviving Mayan book, and the dates inscribed on stone monuments to rulers, are enough to prove that Mayan and other literate pre-Columbian Mesoamerican societies achieved a sophisticated understanding of star and planet movements, and had the ability to engineer buildings oriented to these phenomena (as are the Giza pyramids in Egypt). The Maya wanted to record dates in histories of their kingdoms and great leaders, and to calculate fate according to conjunctions of stars and planets, noting which were in ascendance when a person was born—that is, astrology.[1] You can visit a Mayan astronomical observatory, called the Caracol, in Chichén Itzá in Yucatán, Mexico, and one built by a different Mesoamerican nation at Monte Albán in Oaxaca, Mexico (Building J). In the buildings you will see the oddly spaced openings their astronomers looked through to pinpoint when a star or planet reached a certain point in its annual movement, or when the sun reached its zenith on the solstice or equinox, marking those days.

---

[1] Do not confuse *astronomy*, the scientific observation of sky phenomena, with *astrology*, the inference of fate believed influenced by what was overhead in the night sky when an individual was born. See Chapter 7, "Calendar Astrology," on the elaborate system of sky knowledge, calendar, and supposed personal character traits and fate invented some three thousand years ago, spread as a kind of serious magic to all the major nations of Eurasia and Mesoamerica, and still today seen by millions as the horoscopes printed in daily newspapers.

While the magnitude of mental feats performed by Mesoamerican astronomers was being deciphered, alignments between prehistoric structures and astronomical phenomena were noted in several other parts of the world. Controversy developed between astronomers pointing out the alignments and archaeologists who couldn't believe nonliterate people could make such alignments, since many required recording observed positions year after year over cycles of movement. Stonehenge in England, the best known, caught the worst flak.

Built and rebuilt with changes from about 2900 B.C.E. to about 1650 B.C.E., Stonehenge began as a timber structure surrounded by an earth bank and ditch, approximately 300 feet (100 meters) in diameter. Several centuries later, eighty tall, heavy slabs of a bluish-gray stone were transported from hills 134 miles (215 kilometers) west and stood up in the form of two concentric horseshoes at Stonehenge (Figure 8.1); a raised road, perhaps for processions, was built to the structure. There are several other impressive structures near Stonehenge contemporary with its earlier phase, 2900 to 2600 B.C.E., including the even larger Avebury (a whole village is inside its circle of huge upright stones), Silbury Hill which is the largest earth mound in western Europe (130 feet [40 meters] high, conical in shape), and Durrington Walls which had two circular structures of large timbers rather than stone slabs. The bluestones were subsequently taken down and reset around an inner horseshoe of huge slabs of sandstone from a nearer source, with five of the slabs somehow laid on top of the massive uprights. There is no evidence that this oversized structure ever had solid walls or a roof. Close to the stone circles, another big slab, the "Heel Stone," was set which lines up with the summer solstice sunrise and winter solstice sunset (the longest and shortest days of the year), the sun throwing a beam illuminating the central area within the circle at summer sunrise.

FIGURE 8.1 Stonehenge, England, ca. 1650 B.C.E. *Photo M. Allen.*

At Stonehenge, it is clear that the rectangular "Heel Stone" marks the solstices. Whether the *purpose* of erecting Stonehenge was to note the solstices is another matter; aligning the Heel Stone with the sun's beam shining through at midwinter and midsummer rituals may have been done for dramatic effect. Mind, we don't know that the Britons in the third millennium B.C.E. had solstice rituals, only that historically, festivals have been held on these days throughout Europe and in most societies in latitudes where winter and summer contrast. While this interpretation is often repeated, a disquieting fact is that the solstice correlation so obvious today is now closer to the alignment of the Heel Stone than it was in 2500 B.C.E.

In the 1960s, an astronomer proposed that Stonehenge records not only solstice sunrise and sunset points but also a means of predicting moon eclipses through sets of holes (or stakes in the holes). Predicting moon eclipses rests on an 18.61-year cycle for the moon to rise at a given point on a solar calendar date. A problem with calculating when an eclipse will occur is that one will be visible *somewhere* from earth every 173.5 days, but *where* on earth people can see it changes radically, and of course cloudy weather can hide it. Given that the ancient astronomers had to accommodate a sort of leap year to handle the two-thirds year added to eighteen years for the cycle, and considering how many generations of observers would be required to record so long a cycle, including interference from bad weather, moon eclipse prediction from Stonehenge is much less probable than solstice alignments. Arguments for moon observations did provoke archaeologists to realize that every 9.3 years, a full moon rose over the Heel Stone at winter solstice. That would have looked dramatic in the winter night.

Moose Mountain

A flurry of debates over Stonehenge and over calculations of geometries and astronomical alignments at a number of other megalith ("giant stone") constructions in Europe led to the creation of the subfield of astronomy now called archaeoastronomy. Your author participated in one of the pioneer projects of this new field. In 1974, an astronomer on a family camping trip in the Big Horn Mountains of northern Wyoming noticed that lines of stones in an enigmatic prehistoric construction there could have been used to determine the solstice. The astronomer wrote up his observations, drawing the sightlines on a map of the "medicine wheel" construction, and sent it for publication in the journal *Science*. The editors of the journal asked a Harvard archaeology professor to evaluate the paper, and he, realizing his expertise on Mayan sites did not stretch to Northern Plains archaeology, remembered his former students working in that area, and passed the paper on to my husband and me. I'll always remember Tom coming into the kitchen where I was fixing supper, waving the paper and pointing, "This diagram of the astronomy sightlines looks like our map of the Moose Mountain 'medicine wheel'!" (Figure 8.2). Dinner was over quickly

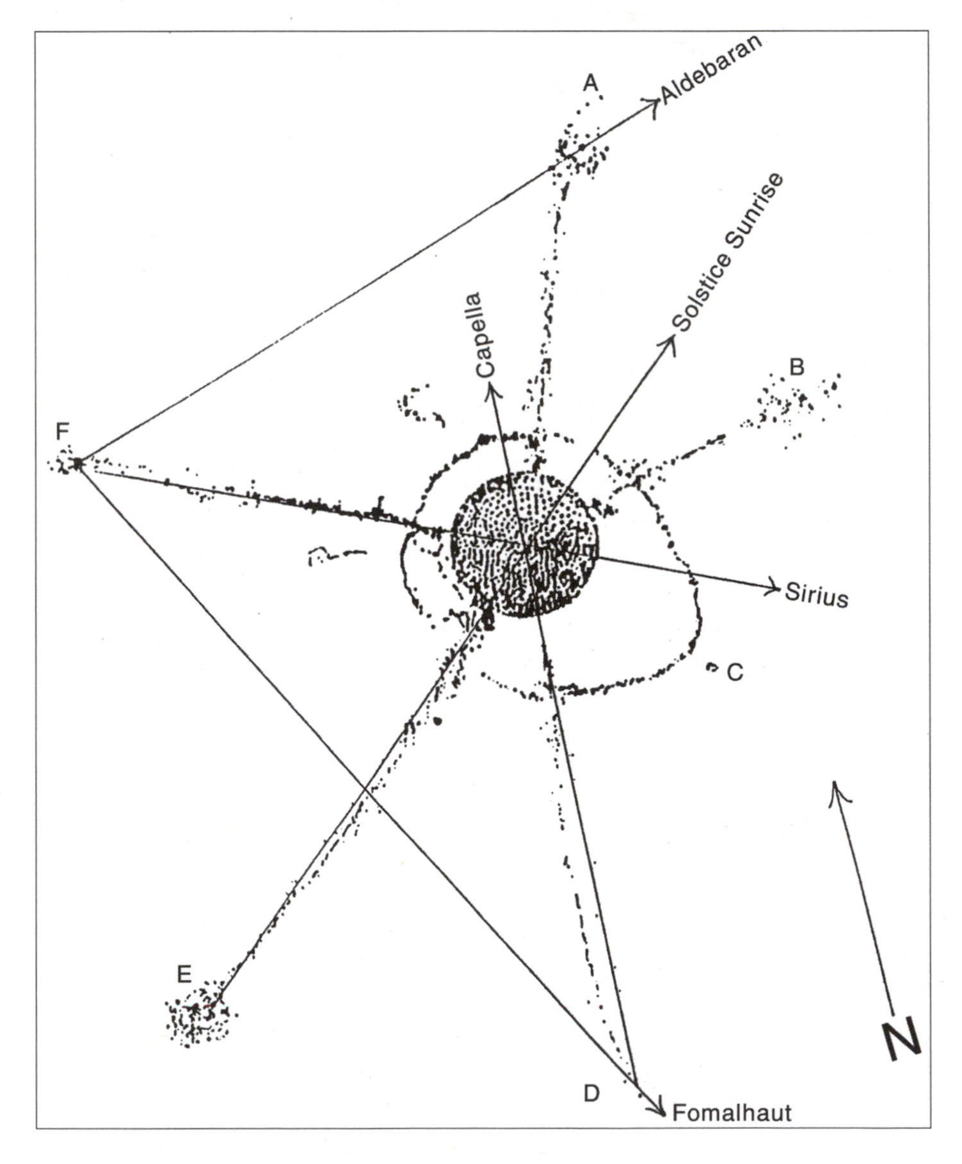

FIGURE 8.2 Map of Moose Mountain medicine wheel, showing sighting lines to summer solstice sunrise and to bright stars. *Map drawn by Thomas F. and Alice B. Kehoe; alignments by John A. Eddy.*

so we could sit down with the map we had prepared years before of a puzzling construction of boulders on the top of a high hill in southern Saskatchewan. About a dozen such carefully arranged circles plus lines of small boulders, usually with a central cairn (heap of stones; see Figure 8.3), were known to archaeologists in the Northern Plains on both sides of the U.S.-Canadian border, east of the Rockies. A few were said, by local Indian people, to mark graves of respected leaders, but others were as strange to area Indians as to us. We urged *Science* to publish the astronomer's paper, which it did, and we sent a copy of the Moose Mountain map to the astronomer. He phoned back, "Can I meet you next June at summer solstice at the site?"

That June, we camped at the base of the hill to assist the astronomer with measuring distances and angles between stone lines and cairns. We got up before dawn, climbing the hill and positioning ourselves at the rock line pointing northeast to

FIGURE 8.3 Central cairn, Moose Mountain medicine wheel, Saskatchewan. *Photo by Alice B. Kehoe.*

await sunrise. Well, June is a rainy month on the Northern Plains. The astronomer was using a long weekend of vacation time (his supervisor didn't think measuring prehistoric stone lines was part of his job description), and each morning was cloudy, rainy, or foggy. One foggy morning, after the totally obscured sunrise had come and gone, we traipsed around and around in the murk before we could see our tent trailer at the base of the hill! After four days, the disappointed astronomer reluctantly had to go back to Colorado. The fifth morning was beautifully clear, and Tom and I, standing along the northeastward alignment, saw the sun come up as predicted, over the central cairn. Only, not over its center, but over its east edge. Was the east edge what the Indians had used to mark the solstice?

Our astronomer collaborator had a challenging alternative hypothesis: Moose Mountain construction was two thousand years old. That long ago, due to wobbling of the earth's axis over millennia, the sun would have appeared directly over the center of the central cairn, and furthermore, five other sight-lines to bright stars would have worked. Two thousand years old? It was hard to believe that these small boulders lay on the surface of the top of a windy hill for that long. Would not soil have blown and mounded up over them? Could the site really be this ancient?

To find out, the next summer we obtained permission for limited excavation on the site from the Indian band owning the hill, and cut narrow trenches along several alignments and through the central cairn. The first thing we noticed was that the lines of visible boulders rested securely on a lower level of stones now buried. Winds over centuries had covered all but the rounded tops of the upper stones. Second, we saw the stone alignments were carefully constructed to be permanent. Third, the central cairn proved to have a lower mound of stones packed tightly with earth, a big white rock balanced on top of the middle of that mound, and an accumulation of hundreds of loose stones piled around the basic construction. We inferred that, after knowledge of the purpose of the construction had been forgotten in later centuries, passersby had put rocks on the cairn just for luck, obscuring the crucial white sighting stone standing on the original mound. An avocational archaeoastronomer, using a computer, confirmed our inference. We recovered little chips of charcoal, enough for radiocarbon dating, under the lowest stones in the central cairn. Bits of red ocher with the charcoal suggested a ceremonial fire had been lit on the cleared earth before the construction of the cairn had begun. The radiocarbon lab upheld the astronomer's calculation of probable date: the Moose Mountain "medicine wheel" dated to mid-first millennium B.C.E. All six of the celestial sightlines would have worked at that time.

Moose Mountain is among the oldest dated astronomical observatories in the Americas, older than the Caracol in Chichén Itzá or Building J at Monte Albán, mentioned above. However, those buildings are final versions of observatories no longer surviving, where records and calculations were developed that enabled the sophisticated structures we can see today. An interesting aspect of Moose Mountain is that it has no connection to marking seasons for agriculture, or to the kind of year-round series of rituals celebrated in urban societies such as in Mesoamerica or the Southwestern Pueblos. Saskatchewan's prairies are too far north to grow indigenous American crops. Since Paleoindian times, hunting bison was the basis of human subsistence there, and people lived in seasonally moved camps. Why would they want to accurately note the solstice? Why would it be so important to them that they made multiple alignments: to the summer solstice itself, to star risings a month before, a month after, and to the evening of the summer solstice, in case solstice mornings were obscured by rain or fog, as they were when we first investigated the hill?

We hypothesize that the bison hunters' small bands of a hundred or so persons gathered annually in July in a rendezvous of several thousand, to trade, celebrate major rituals, work out feuds and make alliances, obtain spouses, and party. Without friends made at rendezvous, known for all the historic Plains Indian nations, small bands would be vulnerable to local disasters and their young people might not find someone to marry. Noting the solstice would have informed the bands that it was time to start traveling to rendezvous.

Part of our investigation of Moose Mountain included interviewing Saskatchewan First Nations people about the site, and about calendar observations. All of the First Nations—Cree, Assiniboine, Saulteaux, Dakota, and Blackfoot—told us that the elaborate construction on Moose Mountain is older than their peoples' occupation of the Saskatchewan prairie. We learned that Cree bands used to appoint a member to be calendar keeper: he had two bags and 365 little sticks, and on summer solstice, one bag was empty and the other full (the summer solstice was their New Year). Every day, the calendar keeper moved one stick to the other bag. In this way, they knew how many days had elapsed since the previous solstice, and how long until the next. If at solstice, one bag was not quite empty—the keeper had somehow skipped a day—the count could be rectified and the new year begun with the proper one empty, one full bag. We asked our Cree collaborator whether his band still had a calendar keeper. "No," he replied, "when the bank began giving out free calendars, we didn't need to rely on a keeper any more."

PYRAMIDS AND MOUNDS

Astronomical alignments are spectacular when they are built into stupendous monuments. Giza's trio of pharaohs' pyramid tombs in Egypt have been exciting admiration for nearly five thousand years. Khufu's, the largest, is 230 meters (755 feet) per side on its square base and was originally 146 meters (480 feet) high. Like Hopewell earthworks in America, the Egyptian pyramids testify to a sophisticated ability to translate mathematical calculations into engineering. Two sides of Khufu's pyramid run straight to true north, with the intersecting sides at a right angle, and the other pyramids and the Sphinx Temple are aligned by their corners to a northeast-southwest diagonal in such a way that the equinox sunset, in spring and autumn midway between solstices, illuminated a sanctuary and merged shadows of the Sphinx and Khafre's pyramid (Figure 8.4). When it came to building from the plans, the pharaohs commandeered hundreds of workers to cut stone blocks, transport them (on sleds pulled over rollers?), and pile them layer by ever-smaller layer until the last made the apex of a triangle. How such

FIGURE 8.4 Pyramid of Giza and Sphinx, Egypt. *Courtesy Milwaukee Public Museum.*

heavy stone blocks were lifted to their positions has intrigued engineers today: were ramps of earth built to haul the blocks up? inclined scaffolding? counterweights? It's hard for us to comprehend the sheer brute muscle, human and ox, a pharaoh could bring to bear on moving tons of stone. And although he controlled such masses of man and beast power, he feared robbers would sneak into his tomb and destroy the body he hoped to be resurrected. Inside the Giza pyramids are passageways and rooms artfully designed to block and confuse those dastardly robbers, more testimony to the engineers' ingenuity.

Pyramid-shaped structures are found in many parts of the world, with a variety of building techniques and functions. In the Americas, pyramidal structures were constructed by raising a core of rough stones and earth, or adobe bricks, or packed earth with drainage layers, and covering it with a veneer of finely dressed stone slabs or of clay. These various methods are also seen in Egypt, a difference being that the great Giza pyramids of the mid-third millennium B.C.E. were finished with finely dressed and tightly fitted blocks of limestone to make a smooth face, instead of slabs. Comparable in size to the Giza pyramids, the Pyramid of the Sun at Teotihuacán in Mexico, built two thousand years ago, is 210 meters (689 feet) at its square base (similar to Khufu's pyramid) and 64 meters (210 feet) high (Figure 8.5), but unlike the Giza pyramids, those at Teotihuacán and elsewhere in the Americas were flat on top, crowned with temples or palaces. Many did contain rulers' tombs, linking the dynasty to the gods worshipped above them on the platform top. Engineers in these American societies were challenged not only by the weight of material to be lifted, but also by the necessity of figuring how to drain the layered mass and how to stabilize it. At the Andean capital city of Tiwanaku, in Bolivia, the principal pyramid has channels through which water flowed onto its terraces, making it into a magic mountain out of which came prayed-for rain. Cahokia's principal platform mound, across the river from St. Louis, Missouri, is a rectangle 316 meters (1037 feet) long, north-south, by 241 meters (790 feet) wide east-west, 30 meters (100 feet) high. A massive timber hall, and other buildings, were on the football-field-sized top. Testing through the mound revealed layers of gravel and clay placed to drain even the very heavy thunderstorms engendered in the region, at the confluence of the Missouri and Mississippi Rivers. The great plaza in front of this mound, a third of a mile long, was built with a thick layer of sandy loam so level that still today, water does not puddle on it. Archaeologists working in this plaza noticed its perfect acoustics for an audience gathered to hear rituals performed on the principal mound. Only slightly affected by recent highway heavy machinery running at its base, this mound, the plaza, and over

FIGURE 8.5 Teotihuacán, Mexico, ca. A.D. 500. *Photo by Alice B. Kehoe.*

a hundred other mounds at Cahokia have retained their shape for a thousand years, an architectural and engineering feat (Figure 3.3, page 66).

From England's largest mound, Silbury Hill (160 meters [525 feet] in diameter, 40 meters [130 feet] high), to its contemporary the far greater Giza pyramids, and American platform mounds, some equally old, the marvel is not that manpower piled up the earth or stones, but that engineering intelligence calculated so well, making these monuments permanent on the landscape for millennia. Why did people build them? An obvious reply is, to impress other people. We ourselves are greatly impressed. The best modern comparison is to the Mall in Washington, DC, an enormous plaza lined with massive, gleaming stone buildings culminating in the Capitol. The hearts of capital cities, like the Mall in Washington, embody the grandeur and power of the nation. They trumpet the message that the government is bigger, and commands more force, than any ordinary organization. Louis XIV, the "Sun King" of France (1634–1715), had the garden of his palace at Versailles laid out to lead the eye to the horizon, conveying the feeling that his majesty extended to the end of the earth. Great cathedrals and temples are on the scale of the glory of God, hopefully attracting God's favor as well as mortals' obeisance. Until the twentieth century, all of these super-scaled constructions were built without powered machinery. Time is the primary difference between ancient and modern technologies, with the ancients accounting for time-consuming hand labor by using large labor forces marshaled by the ruling elite. In modern capitalist societies, machine production is substituted for this mass of human labor.

How did the laborers fare around the pyramids? Were they slaves doing back-breaking work until they died? Excavations carried out in Egypt near the Giza pyramids, and on a site dated a thousand years later near the equally elaborate pharaonic tombs of the Valley of the Kings, reveal villages of small mud-brick homes, a few mansions with gardens for estate-owning aristocrats and top officials, and commercial-size bread bakeries, beer breweries, weaving and carpentry shops, and granaries. Records kept by professional scribes for government and private enterprises, orders, and personal letters tell us that the several thousand men employed building a pyramid were paid with bread and beer—the bread was heavy, whole-grain loaves, the beer rather thick and pretty nutritious—and excavations show local catfish and dried fish were issued, too. The men's families were assisted with rations. Women served in the temples and worked as house servants, weavers, grain grinders, and farmhands, although many married women apparently stayed home to care for children and small livestock, and to cook and make clothing. They prepared meals for their men who were out working, the food carried to the workmen by youths employed by the management. Workers were supposed to have a day of rest at the end of their ten-day week, they could take personal days off when a family member died, and they could and did go on strike if payday was delayed. On construction jobs, men were efficiently organized into work gangs and sets of work gangs. Their work was physically very heavy—for example, dressing stone blocks by pounding relentlessly with stone hammers held with both hands—but the men were not slaves. Instead, laborers' stints at the pyramids took place only part of the year and were a form of taxation owed to the state, while skilled craftspeople and well-paid officials held permanent jobs.

Since we lack texts for Late Neolithic and Bronze Age mound and megalith builders in Europe and mound builders in North America, we don't know historical details in the same way as we do for Egypt. But excavations of settlements near European and American massive mounds indicate workshops and homes for both upper-class and commoner, with farms beyond providing food transported by boat or porters. "Rome wasn't built in a day" applies to these ancient capitals and holy precincts: cutting, moving, and raising the more than two million blocks of stone in Khufu's pyramid could last a generation, and the same for carrying millions of basketloads of soil and clay for Cahokia's mounds and plazas. The beautiful medieval cathedrals in European cities, contemporary with Cahokia, also took generations to build.

■ ■ ■ Superhuman Figures

The use of geometry applied to more than just spectacular pyramids. In Ohio, people made geometric outlines, and in Peru, animal figures, on a scale best seen from a helicopter. Only, they didn't have helicopters. How these unrelated peoples, far separated in space and in time, worked out their constructions by doing geometry on the ground is fairly well understood, but why they chose to work on a superhuman scale isn't known.

Hopewell, the Middle Woodland occupations, ca. 100 B.C.E.–A.D. 400, of Ohio, built amazingly precise, gigantic geometric figures of no obvious practical use (Figure 8.6). Ritual or ideological features include constructing walls of two parallel colors of soil, the lighter facing inward to the site, the darker facing outward, and then covering these red-and-yellow walls with ordinary brown soil to complete the desired height, in some instances several meters high. Geometric figures may be circles, octagons, squares, or rounded-corner rectangles, encompassing areas ranging from a few meters to the Newark Octagon enclosing an entire modern golf course including clubhouse. Mounds of various size may be inside the earth-wall enclosures, and may hold burials, sometimes in log tombs containing sacrificed retainers and precious objects around a central personage. Earthworks in several Ohio valleys lie in pairs, usually a circle and a polygon, often with straight parallel walls connecting them or built causeways leading to them; portions of a straight built road 90 kilometers (60 miles) long between the great multi-figure earthworks at Newark (Ohio) and the High Bank site at Chillicothe, Ohio, have been plotted. Newark and High Bank each have a circle 320 meters in diameter—yes, identical although 90 kilometers apart! Octagons and construction with two-color walls inside final embankments also pair these sites. Within a single valley, sites often pair together, in some instances as mirror images, and angles of figures often point to figures in sites kilometers away, suggesting overall master plans, or at least conscious linkages, for entire river valleys.

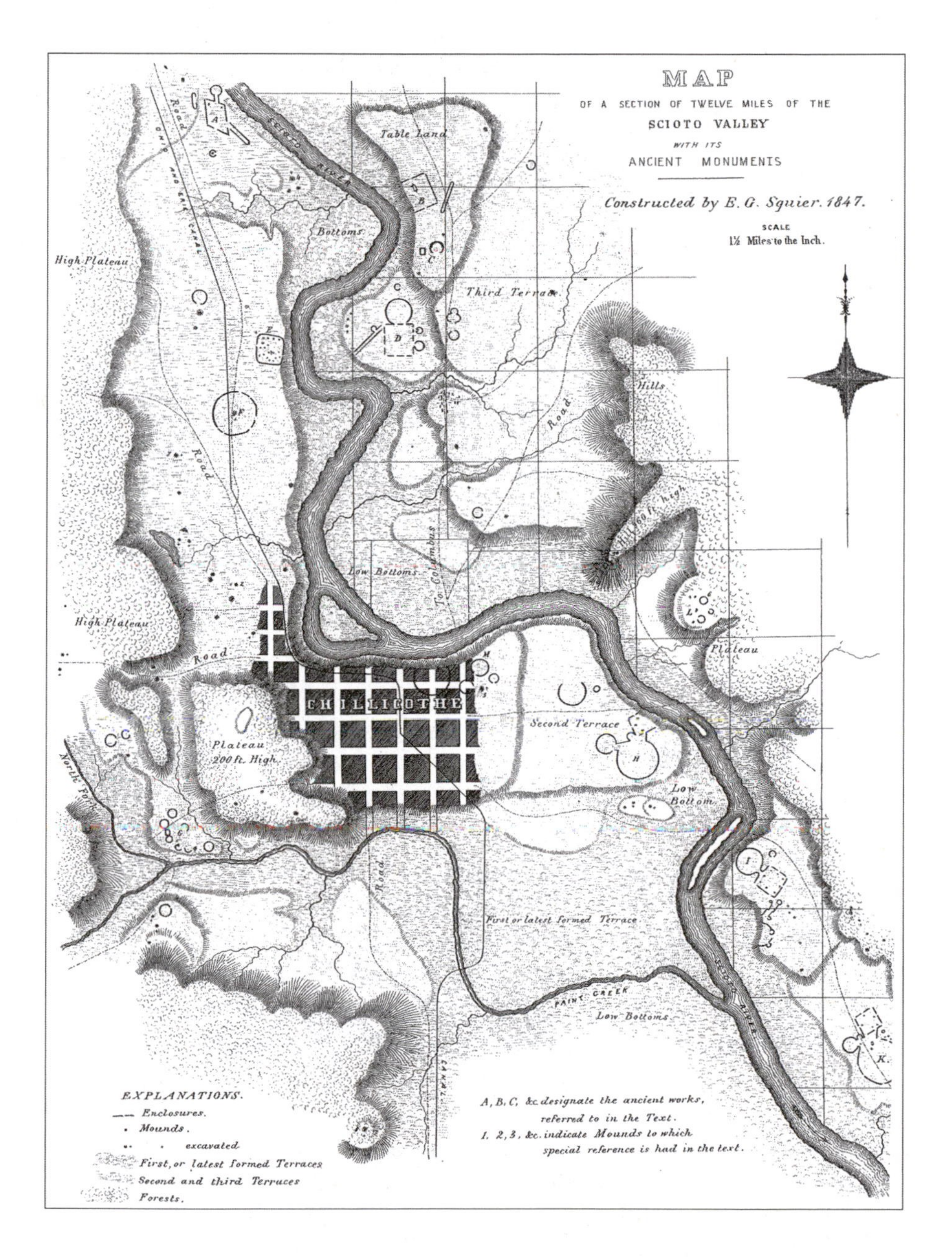

FIGURE 8.6 Map of Hopewell geometric earthworks, ca. A.D. 200, around Chillicothe, Ohio. *Drawn by E. Squier and E. Davis, 1848. Smithsonian Institution.*

What blows our minds today is that Hopewell people lived in hamlets of small pole-and-thatch homes, cultivating semi-domesticated local plants (goosefoot [*Chenopodium*], knotweed, some sunflower for its seeds) and a very small amount of maize. They had no cities, no fortresses, no writing so far as we have discovered. Yet their engineers directed dozens, possibly hundreds, of laborers to impose on the landscape giant figures that Euclid the mathematician might have dreamed of, but his Greek fellow-citizens did not attempt. Hopewell leaders purchased rare raw materials such as black obsidian (volcanic glass) quarried at Yellowstone Park in the Rocky Mountains two thousand miles to the west, fragile mica sheets from the Appalachians to the south and east, copper from Lake Superior to the north, and shells from the distant Gulf of Mexico. They commissioned trained artisans to work these rare imported materials into symbols of power—hawks, upraised hand, stag antlers, bears, trophy heads. Fine textiles wrapped around copper objects testify to spinning and weaving arts; contact with copper preserved these textiles which otherwise would have decayed away, leaving no evidence of these arts. Astronomical orientations of some of the earthworks demonstrate astronomy advanced enough to pinpoint and mark, at Newark and High Bank, moon-rise maximum and minimum positions, north and south.

Archaeological evidence has been consistent that the Hopewell lived in small groups scattered across the landscape, harvested a variety of local resources, cultivated indigenous plants, hunted game and caught fish. Controversy arises over whether a subsistence economy like this can produce such elaborate architecture, contrary to Western ideas of cultural-evolutionary stages. Hopewell culture reveals the racist tradition behind common textbook formulations of a straight path from hunting-gathering to village agriculture to complex urban societies. Evolution occurs in a multitude of pathways. The Hopewell lived comfortably in Ohio's rich valleys, as their ancestors had lived for millennia. That did not hinder them from recognizing daring intellectuals and arrogant warlords who probably planned and ordered these elaborate earthworks. There shouldn't be controversy over the civilized achievements of such nations who relied on diversity of natural resources rather than imposed foreign crops. Why they stopped creating extraordinary geometric earthworks after several centuries isn't understood. It may be that bows and arrows, weaponry newly introduced about the time Hopewell ends, made it dangerous to reside dispersed over the landscape. Profound knowledge of geometry is no defense against an arrow whizzing out from the forest, beyond range of your own soldiers armed with spears and darts.

Easter Island

"Mysteries of Easter Island" is a favorite popular trope. Are there mysteries? Archaeologists like to say, "We don't really know . . . " because the more data one gathers, the more one realizes gaps remain.

Where did the Rapa Nui (the islanders) come from? "We don't really know . . ." except it had to be the Marquesas and/or Mangareva/Pitcairn islands in the mid-Pacific.

How many people lived on the island when rock-mulch fields were built? "We don't really know . . . ," but perhaps up to ten or twelve thousand.

How were the giant statues (Figure 8.7) transported? "We don't really know . . . ," but probably they were skidded up over the lip of the crater by dozens of men pulling ropes, slid to their platforms along prepared roads using log rollers (according to one local legend, greased with mashed yams and sweet potatoes), and levered up onto the platform with crowbars—all methods experimentally validated by recent researchers.

Easter Island promises mystery because it is situated so incredibly far from anywhere. Mysteries dissolve into reasonable research projects once Easter Island is seen as part of history's greatest colonization program and comfortably within the general context of Polynesian culture.

FIGURE 8.7 View of Easter Island. *Photo by Christopher Stevenson.*

Rapa Nui, called Easter Island by its European "discoverer," was colonized by Polynesians about eight hundred years ago. The colony sailed from Central Polynesia, perhaps from the Marquesas via Mangareva or Pitcairn Island, perhaps directly from the Marquesas. The first controversy is over whether their settlement was planned after an exploratory voyage reported the island, or was lucky chance. Big double canoes loaded with men and women, chickens, semi-domesticated Polynesian rats, taro, yams, bananas, sugar cane, paper mulberry trees, and probably sweet potato, not to mention dried fish, drinking water, tools, and clothes, did land after sailing a little under a month from Mangareva or Pitcairn, to judge from a replica voyage in 1999 by the Hawai'ian-built sailing canoe *Hokule'a*. Rapa Nui traditional history says a chief sent out a six-man exploratory canoe, they found Rapa Nui, presumably made their way back against prevailing winds or during an El Niño period of westerly winds, and led the settlers to the island. It's worth noting that although Rapa Nui is a speck of land in a huge open ocean, its presence could be guessed by Polynesian seafarers when they saw seabirds flying around, a few days' sailing from the island itself. To get back, the explorers would follow stars they had noted as they sailed, the Polynesian "star compass" they kept in their heads and modeled with shells tied to networks of sticks as a map of the night sky.

Settlers found a volcanic island of rocky ocean cliffs and an inland crater with a lake in the center. Palm forests, bearing nuts like little coconuts, covered much of the land. The Rapa Nui cleared fields for agriculture and small villages, ingeniously adapting methods to the colder, less tropical climate than that of their homeland. For several centuries, the Rapa Nui prospered, keeping their bananas and sugar cane in sheltered, warmer locations on the island. By about A.D. 1400, so much of the island was occupied that people began inhabiting poorer areas without springs or seasonal streams, guarding against too-rapid evaporation of rainfall by covering fields with a mulch of rocks. The same technique was used by the Hohokam in arid Arizona in the U.S. Southwest. Layers of rocks allow precipitation to reach the soil, but screen against evaporation. Meanwhile, the little, mouse-size Polynesian rats enjoyed for food by aristocratic households escaped their cages and proliferated. They gnawed the nuts (that is, the seeds) of the palm trees, and ate birds' eggs, preying on the guano-producing birds hunted by Rapa Nui humans. A number of bird species disappeared, and the slow-growing palms (related to Chilean wine palms) failed, unable to regenerate. Lacking big logs, the Rapa Nui could no longer build seagoing canoes, but had to make do with pathetic little ones patched out of short pieces of wood; and lacking seagoing canoes, men no longer brought in tuna, dolphins, and other large marine game. With more and more people struggling to survive with declining resources, warfare erupted. That, of course, isn't a creative way to deal with crises: the Rapa Nui began to carve huge statues out of the volcanic rock in crater slopes and set them up on platforms (low versions of platform mounds). They built roads to transport them from the quarries to their platforms, ringing the coast at good harbors and habitation locations. The statues represent deities; as stress worsened in the overpopulated island, aristocrats commissioned these tributes to their gods, raising them where community priests could invoke their favor. Unfortunately, neither labor-intensive rock-mulch agriculture nor imposing statues of deities could make everyone happy. By about 1680, food shortages, lack of wood, and frequent battles plagued Rapa Nui. European contacts sporadically in the eighteenth century (more often than recorded, as whaling ships stopped by without leaving historical notes) must have been another wrench in the creaking machinery of traditional Polynesian society. Apparently in the 1770s, some Rapa Nui began to pull down the giant statues, not all at once but over the ensuing century that saw increasing foreign depredations, including hauling islanders off to South America as slaves.

The island of Rapa Nui shows in its small space many marvelous technical accomplishments. Without any iron, the people of Rapa Nui constructed ships to carry tons of cargo a thousand miles or more, and sailed precise, targeted routes without compasses or GPS units. They had to figure out how to protect their tropical crops from their new island's colder climate, and when they needed to expand agriculture to feed their increasing population, they ingeniously constructed rock mulch on the drier soils. They used stone picks and adzes to quarry stone for tall statues and sheer massed muscle, with rollers, to move the giant figures. Patient hard work coupled with intelligent analyses of problems have moved mountains—or at least huge slabs of them to Stonehenge and on Rapa Nui—and in the world's pyramids and mounds, created man-made mountains. Technology is not a modern, nor a Western, invention; it is a human gift that we should marvel at.

■ ■ ■ PEOPLING OF THE AMERICAS

Humans coming into the Americas before the Pleistocene Ice Age had ended reveal, if you think about it, impressive evidence of advanced technologies overcoming formidable geographic obstacles. Humans did not evolve in the Americas—our continents broke off from the Eurasian-African land mass before any ape-type primates came about, much less humans—so we look for evidence of the first humans to colonize from Asia, where populations most like American Indians developed toward the end of the Pleistocene, around ten thousand years ago. Everyone agrees some early migrants came through the Bering Strait region where Siberia is only a hundred miles from Alaska.

Disagreements arise over when early migrants came, and whether they used boats or walked across a Pleistocene land bridge later flooded when the continental glaciers melted. From time to time, someone argues that in addition to Asians coming into Alaska, ancient Europeans might have moved along the islands of the North Atlantic down into eastern Canada. A few mavericks point to the narrowest part of the Atlantic, between West Africa and Brazil, and postulate dugout canoes carrying late-Pleistocene settlers from Africa.

As mentioned in the previous chapter, it must be accepted that water transport was invented by 50,000 years ago, because people crossed open ocean between New Guinea and Australia then, attested by a few human remains of that age found in Australia. Whether these pioneer Australians used rafts or canoes, we do not know. With people in southernmost Asia making watercraft, eventually moving into the nearer Pacific islands and onward, it is likely that people in northern Asia also knew how to build rafts or boats, raising the possibility that some people paddled along the coasts of the North Pacific and eventually, over generations, down the western coast of America all the way to Chile. We have, so far, no actual remains of Pleistocene boats. The oldest recovered boats are dugouts sunk in muck a couple thousand years after the end of the Pleistocene. Boats that might have been used to coast the North Pacific or North Atlantic in the late Pleistocene presumably would have been hide-covered frames like Inuit umiaks and British curraghs (similar to birchbark canoes), because trees big enough for log dugouts would not have grown in those far northern lands.

Controversies rage over:

- whether humans colonized the Americas before a specific stone-working technology called Clovis appeared (called the Clovis First debate);
- whether colonizers came only over the Bering Strait/Beringia;
- whether there was more or less a single continuing movement south along the Pacific coast;
- whether any colonization came from western Europe, and if so, who and when; and
- how early the migrations began.

Nineteenth-century archaeologists struggled to estimate ages of American prehistoric sites and artifacts. Without radiocarbon-isotope and other laboratory technologies to calculate years elapsed, some mistook roughed-out quarry stones for million-year-old Early Paleolithic artifacts; others went to the opposite extreme, insisting that sites and artifacts couldn't be more than two thousand years old. The breakthrough came in 1927 when excavation of a New

Mexico site from which bison bones were eroding revealed a finely crafted stone weapon point between bison ribs. Calls went out to a paleontologist at the American Museum in New York, an archaeologist at the Smithsonian in Washington, and the most respected Southwestern archaeologist. The three authorities agreed that the large bison bones belonged to a species that became extinct about ten thousand years ago, and that the stone artifacts had been used to kill and butcher the bison. Therefore, there were people in America that long ago, hunting huge animals soon to become extinct.

Soon after "Folsom points," named after the village near the site with bison bones, became accepted as evidence of late-Pleistocene migration of humans into America, a somewhat similar style of stone weapon blade was discovered in another New Mexico site, near the town of Clovis. This type of stone blade is larger than the Folsom type, and the characteristic long, vertical hafting groove extends only about a third of the way up the blade, whereas the Folsom type has the groove nearly up to the tip. The two types share virtuoso flint-knapping (chipping) skill, making the artifacts beautiful and easily noticed. At Blackwater Draw near Clovis, excavation demonstrated that Clovis points were in a layer lower than the layer with Folsom, indicating that Clovis technology was older than Folsom. Clovis but not Folsom points have been found with mammoths, another clue to Folsom being later, after mammoths became extinct in America. Clovis-style points have been found over most of North America, even more in the East than in the West. Proponents of the Clovis First theory claim that this style of blade was ideally suited for hunting mammoths and other large animals and was used by the first wave of migrants moving south into the unpopulated Americas from Beringia. Opponents to the Clovis First hypothesis interpret Clovis as a technology and style that spread through already-populated North America. It was, you might say, the Coca-Cola of terminal Pleistocene, a readily identified item diffusing rapidly and widely.

The answer to the debate hinges on whether there is any evidence of settlers in the Americas before Clovis. Clovis First archaeologists demanded copious evidence for earlier settlers, against those thinking there is an uncountable number of sites destroyed or not yet discovered. Lack of evidence in archaeology should not be construed to prove that something never existed. Evidence for humans in America earlier than the Clovis horizon is thin—thin, but there. Skeptical archaeologists sticking to carefully excavated and recorded, abundant data say sites alleged to predate Clovis may be incorrectly dated, or the simpler artifacts recovered might come from camps used by Clovis-equipped hunters who took their weapons with them when they moved on—neglecting to mark their sites for archaeologists way off in the future. Opponents to "Clovis was the first" say, of course, that the evidence is thin, that there have been many more years for sites to be buried or destroyed, and furthermore, first colonizers would have been less numerous than their descendants, so their sites would be less numerous. Most important, we don't have a clear idea of what pre-Clovis technology looked like. What do you look for to find a pre-Clovis settlement?

A newer twist in the Clovis First debate is the assertion that Clovis technology looks like the stone tools of Eurasian Upper Paleolithic Solutrean culture, most famous for Paleolithic cave art. This culture predates Clovis by several thousand years. Proponents of the Solutrean theory see hunters paddling along the North Atlantic ice front spearing seals and fish, and then settling throughout eastern North America, modifying their tool technology eventually into Clovis blades that would become the must-have fashion throughout North America around 11,000 B.C.E. A key part of the Solutreans-become-Clovis postulate is a similarity in stone technology between the two: the "overshot" flint-knapping technique occasionally resulting in a hinge-like fracture along one edge of a blade. The technique is so distinct from pre-Clovis stoneworking, say the proponents, that it would be unlikely for earlier Americans to have independently invented the technology; they use Ockham's Razor (Chapter 2) in pointing to an already-existing practice, albeit one fairly distant in time as well as place. Flint-knappers neutral to the debate, such as Canadian archaeologist Eugene Gryba, himself experienced in Paleoindian excavations, say that the technique, and producing the hinge fracture, is not especially difficult for a highly skilled knapper. We can emphasize that "Solutrean" encompasses a wide variety of artifacts over several millennia in Europe, and that there are similarities in addition to the virtuoso blades but not a close matching overall in artifact inventories. The gap in time, around five thousand years, between European Solutrean and American Clovis also needs to be better accounted for. At the same time, given our knowledge of sailing technology, we can't discount possible movements from Europe to Canada along the North Atlantic islands, or from Spitsbergen, the westernmost tip of Norway, into northeastern Canada. If sites with the knapping technology in question can be found in these regions, or sites in eastern America with the technology between 16,000 and 11,000 B.C.E., the hypothesized Solutrean connection would be more acceptable.

Researchers claiming to find pre-Clovis sites have produced radiocarbon dates earlier than Clovis. One of the most ballyhooed sites, Meadowcroft Rockshelter in the coal country of northwestern Pennsylvania, contains only Holocene (modern) animals—no caribou, mammoth, or other animals or plants that went extinct at the end of the Ice Age—and an eroding coal seam running along the rocky bluff just over the lip of the rockshelter. Meadowcroft's excavator insists his radiocarbon assayer filtered out all possible coal dust, but suspicion lingers that tiny particles of millions-of-years-old coal would be enough to make the radiocarbon years estimate older than the human occupation actually was. In contrast to Meadowcroft, two sites near Kenosha in southeastern Wisconsin contain disarticulated mammoths with butchering marks on some bones, and in one site, a stone knife blade lying underneath a mammoth pelvis. Radiocarbon dates for these sites, near glacial Lake Michigan and free from any recognized contaminant, put them at 11,500 B.C.E, earlier than Clovis. It is interesting that geologists consulted on interpreting these sites, called "Chesrow complex" occupations, remark that the ice front of the great continental glacier would have been so close that people could have looked up from butchering their mammoths and seen the gleam of the towering wall of ice to the north. Paleoecologists say that melting glaciers deposited a rich accumulation of nutrients, making the tundra in front attract populous herds of mammoths and other prime game—and, the Chesrow sites tell us, human hunters.

Warmer climes probably also had pre-Clovis inhabitants. The Gault site in eastern Texas (Figures 8.8, 8.9), Cactus Hill in Virginia, and Topper in South Carolina are multilevel open-air sites carefully excavated over several years by competent professional archaeologists. Radiocarbon assays date their lower occupations some centuries before Clovis, not a big gap in time earlier, although there might be older occupations not yet exposed by archaeological crews. The pre-Clovis occupations seem to be nested beneath Clovis layers, confirming their older date. It seems easier, to many archaeologists, to see Gault, Cactus Hill, and Topper as evidencing an American population growing from migrations perhaps a millennium or so earlier, then jumping on a bandwagon carrying a technique for making more effective, and strikingly handsome, weapon points—Clovis style. Just as guns spread rapidly around the world, Clovis spearpoints could have been eagerly adopted by hunters seeing how effectively they struck down mammoths. While skeptics challenge how quickly Clovis techniques spread across existing small populations in North America, it strains credulity much more to picture *families* with Clovis spearpoints spreading so rapidly, in a few centuries, across a wholly uninhabited continent.

Monte Verde, in Chile, is a touchstone site in the peopling-of-the-Americas debate. It is almost as far from the Bering Strait as one can get, and its radiocarbon dates of 12,000 B.C.E. make it among the earliest credibly excavated sites. Its hunted game includes mastodon, extinct after 10,000 B.C.E. The site's boggy location preserved organic materials, such as wooden planks apparently used to construct small houses (perhaps covered with mastodon hide), an abundance of gathered plants, and wooden tool handles as well as stone, bone, and mastodon ivory artifacts. Some of the remains indicate the people traveled to the Pacific seacoast west of their valley, bringing back marine products. After several years of heated discussions as to the date of Monte Verde, a blue-ribbon group of Paleoindian archaeologists was taken there to see the site first hand, and to the excavator's lab to examine its artifacts. The committee reached a consensus and published a report accepting a pre-Clovis date for the site. Then a North American archaeologist—not an academic but a person employed in contract archaeology—combed through the massive report on the site, noting many discrepancies. Were they the result of sloppy copy-editing for the book? Or, as the excavator claimed, did they arise from trying to make a coherent story out of years of accumulated detailed site and lab reports? Or, might the excavator, convinced of the validity of his interpretation of his work, have overlooked discrepancies?

FIGURE 8.8 Excavation of an artificial stone floor in Area 12 of the Gault Paleoindian site, Texas. Scattered across its surface and around it were Clovis artifacts, bone, and burnt chips from flint-knapping. Clovis artifacts were thought to be the earliest in the Americas, but at the Gault site, a layer below this Clovis floor revealed an even earlier human settlement that used different artifacts. *Courtesy Gault Project, Texas Archaeological Research Laboratory, University of Texas at Austin.*

FIGURE 8.9 Stone artifacts from the Gault site, Texas, ca. 11,000 B.C.E. *Clockwise from upper left*: A fragment of a resharpened Clovis projectile point, a Clovis prismatic stone blade, and two examples of flakes of non-Clovis technology, found stratigraphically below the Clovis levels. *Courtesy Gault Project, Texas Archaeological Research Laboratory, University of Texas at Austin.*

Most archaeologists accept Monte Verde as an authentic pre-Clovis site, and, because of its early date and location on the Pacific coast of South America, figure it implies Pacific coastal migration. Sites associated with this migration will be hard to find due to post-Pleistocene sea-level rise. As the Pleistocene glaciers melted, so much water poured into the seas that they flooded coastal zones of the late Pleistocene era. Sea level stabilized only five thousand years ago. While the tool repertoire of Monte Verde is unique, two thousand years later, diagnostic-style "fishtail points" were being made in Chile and Patagonia, apparently developed from Clovis flint-knapping technology. Most archaeologists realize that accepting Monte Verde opens the door to accepting an increasing number of sites and remains earlier than Clovis. Without stone artifacts in a distinctive, readily recognizable style or technology, archaeologists can't make a diagnostic set to identify ancient groups of migrants, their movements and settlements. The "Clovis First" controversy rests upon a particular technology that, in the view of some archaeologists, fueled a rapid spread of mammoth hunters over a continent, or in the view of others, was more like the Beatles' music sweeping through receptive American communities.

SOURCES

Aveni, Anthony F., 2001. *Skywatchers.* Austin: University of Texas Press.

Barton, C. Michael, Geoffrey A. Clark, David R. Yesner, and Georges A. Pearson, 2004. *The Settlement of the American Continents.* Tucson: University of Arizona Press.

Charles, Douglas K., and Jane E. Buikstra, editors, 2006. *Recreating Hopewell.* Gainesville: University Press of Florida.

Flenley, John, and Paul G. Bahn, 2003. *The Enigmas of Easter Island.* Oxford: Oxford University Press.

Haynes, Gary, 2002. *The Early Settlement of North America.* Cambridge: Cambridge University Press.

Hornung, Erik, 2001. *The Secret Lore of Eygpt: Its Impact on the West.* Translated by David Lorton. Ithaca, NY: Cornell University Press. (Original: *Das esoterische Aegypten*, 1999, Munich: C. H. Beck'sche.)

Hunt, Terry L., and Carl P. Lipo, 2006. Late Colonization of Easter Island. *Science* 311: 1603–1606.

Jablonski, Nina G., editor, 2002. *The First Americans: The Pleistocene Colonization of the New World.* San Francisco: California Academy of Sciences

Kehoe, Alice B., and Thomas F. Kehoe, 1979. *Solstice-Aligned Boulder Configurations in Saskatchewan.* Canadian Ethnology Service Paper No. 48, Mercury Series. Ottawa: National Museum of Man.

Kelley, David H., and Eugene F. Milone, 2004. *Exploring Ancient Skies: An Encyclopedia of Archaeoastronomy.* New York: Springer.

Kennedy, Roger G., 1994. *Hidden Cities*. New York: Free Press and Penguin.

Krupp, E. C. 1978. *In Search of Ancient Astronomies*. Garden City, NY: Doubleday.

Lavallée, Danièle, 2000. *The First South Americans*. Translated by Paul G. Bahn. Salt Lake City: University of Utah Press.

Lehner, Mark, 1997. *The Complete Pyramids*. London: Thames and Hudson.

Lesko, Leonard H., editor, 1994. *Pharaoh's Workers: The Villagers of Deir el Medina*. Ithaca: Cornell University Press.

Pauketat, Timothy R., 2004. *Ancient Cahokia and the Mississippians*. Cambridge: Cambridge University Press.

Stevenson, Christopher M., Joan Wozniak, and Sonia Haoa, 1999. Prehistoric Agricultural Production on Easter Island (Rapa Nui), Chile. *Antiquity* 73 (282): 801–812.

Straus, Lawrence Guy, 2000. Solutrean Settlement of North America? A Review of Reality. *American Antiquity* 65 (2):219–226 .

Van Tilburg, Jo Anne, 1994. *Easter Island*. Washington, DC: Smithsonian Institution Press.

9 Neandertals, Farmers, Warriors, and Cannibals

Bringing in Biological Data

Biological anthropologists, who study the physical aspects of humans, and archaeologists, who study human material culture, bring complementary experiences and questions to each other's work. Differing standpoints, experience, and methods can, on the other hand, produce controversy. Some disputes are supported by proponents of one side using obsolete information from the other, such as promotion of the reductionist interpretation of rock art being made by "shamans" coming out of trance (Chapter 6). Yet, cautious scholars can illuminate significant differences in basic premises or corollary assumptions underlying controversies and work to resolve them toward a greater understanding of the past.

In this chapter, we begin with a controversy involving both archaeological and biological anthropological data that began over a century ago with the theory that another race had lived in western Eurasia before our ancestors. Another controversy concerns the spread of agriculture and of Indo-European languages; did the two spread widely through migration of a population, or independently through various peoples learning to farm, or learning a new language, perhaps from conquerors? And that brings us to another controversy informed by both biological and archaeological data: when did wars and conquests begin? Is war hard-wired into our primate genetics—like the chimpanzees who attack neighboring groups—or did wars begin only with the development of territorial states based on agricultural economies? A related but dramatic part of the warfare controversy is whether unfortunate people seem to have been not only killed by their enemies, but cut up and eaten as well. Have there been societies that treated other humans as meat? All these arguments require the expertise of biological anthropologists as well as archaeologists.

NEANDERTALS AND US

A long-standing puzzle has been the relationship between Neandertals (*Homo neanderthalensis*) and "anatomically modern humans" (*Homo sapiens sapiens*), our species. On and off, the controversy rages as to whether the two types are more like subspecies or races rather than distinct species—that is, should they be *Homo sapiens neanderthalensis* and *Homo sapiens sapiens*? The difference is more than one of names: if they are subspecies, then they could interbreed. We may have Neandertal genes inside us. Based in part on analysis of remnants of mitochondrial DNA (from mothers) from Neandertal bones, biologists estimate the two types diverged from a common ancestor (*Homo erectus*) nearly 400,000 years ago, supporting the decision to see them as a separate species. African *Homo* populations then evolved, in multiple lineages, into anatomically modern *Homo sapiens* by about 160,000 years ago. Neandertals lived in Europe and western Asia from about 300,000 years ago, adapting physically to the Pleistocene's periodic cold climates in their latitudes. About 100,000 years ago, anatomically modern humans moved eastward from North Africa into western Asia, and encountered Neandertals there.[1] Did this meeting lead to the interbreeding and eventual disappearance of distinct Neandertal characteristics, or was there physical extinction of Neandertals and replacement by *sapiens?* DNA evidence has proven inconclusive. Did our ancestors commit genocide against Neandertals? Anatomically modern humans appear in Europe 40,000 years ago, Neandertals persist to about 28,000 years ago; they occupied the same landscape for over 10,000 years. After that, all of Eurasia and Africa (and Australia) had only anatomically modern people.

Archaeologists interject a question here: Do we assign Middle Paleolithic artifacts to Neandertals, and Upper Paleolithic artifacts exclusively to *sapiens*? Middle Paleolithic technology, dating about 180,000 to 30,000 B.C.E., made stone tools by chipping on round flakes; Upper Paleolithic tools, about 40,000–8500 B.C.E, were more efficient but more difficult to produce. Or did Neandertals learn some techniques from their *sapiens* neighbors and make some of the tools we call Upper Paleolithic? This would imply a greater level of technical skill than has been assumed for Neandertals. Early *sapiens* in Africa, where

[1] By convention, separate species cannot interbreed and produce fertile offspring; horses and donkeys, for instance, can mate but their offspring are infertile mules. Yet biologists sometimes bow to traditional categories and, for example, make wolves, coyotes, and dogs separate species even though their offspring from interbreeding are fertile. In the case of Neandertals and *sapiens*, we have no evidence on the fertility of interbred offspring, if any.

there were no Neandertals, made Middle Paleolithic artifacts comparable to the Eurasian Middle Paleolithic. In other words, for thousands of years, anatomically modern *sapiens* humans in Africa had technology similar to that used by Neandertals, their contemporaries in Europe. The most intriguing situation is in France where artifacts in the Châtelperronian style have been excavated along with Neandertal fossils in a couple of sites. At other sites, with Upper Paleolithic occupations, a scenario suggests that both *sapiens* and Neandertals camped at different times.

From an archaeological standpoint, we can question whether Châtelperronian tools represent a culture distinguishing some ethnic communities of modern *sapiens* from others who preferred to create what we call Upper Paleolithic tools. Is the designation "Châtelperronian" simply an invention of archaeologists sorting out artifacts in the laboratory? Or does the difference represent variances in technology between Neandertals and *sapiens* of the same era? French Paleolithic archaeology is highly scientific, its fine-tuned stratigraphy developed through in-the-field collaboration with sedimentary geologists, but this very sensitivity to the way occupation debris can be amalgamated or filter down may make it more difficult to decide, thirty thousand years later, between brief camping episodes in a site. The difference between Châtelperronian and its successor technology, Aurignacian, which is clearly associated with anatomically modern humans, is that Châtelperronian used a stone-working technology familiar to the Middle Paleolithic, and Aurignacian employed a different knapping procedure.

Aurignacian sites also contain a greater variety of bone, antler, ivory, and shell artifacts, as do later Upper Paleolithic sites, compared with Middle Paleolithic sites. Châtelperronian occupations have some such artifacts not found in earlier Middle Paleolithic assemblages, but they are not as numerous or varied as those commonly found in Aurignacian sites. There is no clear break between European sites that are clearly Middle Paleolithic and Neandertal and those that are Upper Paleolithic and *sapiens*. Châtelperronian sites seem to have elements of each. Archaeologists make fine distinctions as they define artifact styles to distinguish earlier from later people, thus constructing a culture history for a region. How do these distinctions match ancient realities?

Did modern humans and Neandertals interbreed? Are modern Europeans a product of mixed subspecies? Back in the 1930s, anthropologist Carleton Coon remarked that his extensive ethnographic fieldwork in Asia, North Africa, and marginal areas of Europe showed him that whatever the formal relationships between two populations, men will be looking for women for sex. Marriage is one thing, one-night stands quite another. Trobriand Islanders of the South

Pacific explained that sexual intercourse was not necessary for every conception, once "the way was opened" for children's spirits to enter a woman. They pointed to a certain retarded woman shunned by the community who had borne several children although it seemed unbelievable that any man would ever want to sleep with her. Coon pictured a few young *sapiens* men exploring in advance of their band's move into new country. The young guys would see a couple of nubile Neandertal maidens gathering food. By force or persuasion, nine months later, cute little hybrid babies would be born. Maybe, in the best of all possible worlds, the respective grandparents would embrace and live peacefully in neighboring camps. Coon said he could discern a weak expression of some Neandertal traits (stocky body, short extremities, long low skull, deep-set eyes) in his close examination of contemporary populations in areas of Europe with the greatest number of late Middle Paleolithic sites and Neandertal skeletons. Individuals so marked were quite as intelligent as anyone else.

Archaeologist Donald Henry, working with colleagues specializing in studying lithics (stone artifacts), phytoliths (silica casts made by plant roots), chemistry of occupation floor areas, and identification of activity wear patterns on lithics, excavated two occupation floors in a rockshelter in the Middle East, Tor Faraj in Jordan. Dated between 69,000 and 49,000 years ago and containing artifacts normally associated with Neandertal fossils, this site is premised to represent Neandertals, although no human bones were recovered. This was a disappointment, since Neandertal graves—deliberate burials of deceased community members, in a couple of instances with ceremonial objects—have been discovered in a number of Middle Eastern and European cave sites. Henry and his associates observed that along the back wall of the shelter, phytoliths of grasses were concentrated, which they inferred to be from bedding. A series of hearths were aligned with the back wall, about a meter (a yard) out from it, suitable for warming sleepers lying against the back wall. Another series of hearths were situated around two meters (two yards) from the back wall, toward the mouth of the shelter, with primary stone-knapping taking place around these hearths. Stone artifact edges were sharpened here but also nearer the back wall. Butchering of game was mostly done in the back area, by the hearths there, so food may have been cooked in this more sheltered zone. Based on the clustering of hearths and the activities around each, Henry believes the people were organized into families who slept and ate together within the community gathering. Checking where historic Bedouin herders camped in this mountainous region, Henry inferred the Middle Paleolithic occupations were winter settlements, used to maximize warm, dry places for activities and sleeping. Henry concluded that Middle Paleolithic occupants of this cave—that is, Neandertals—were organized similarly to *sapiens* communities engaged in similar activities, and made similar choices for locating their shelters. In short, Neandertals were more like us than not.

From Henry's work, we understand that Neandertals were very human in the manner in which they hunted game and gathered plant foods (dates and pistachio nuts, among others, at Tor Faraj), prepared food and hides, and arranged activity areas in campsites. But the Tor Faraj Middle Paleolithic has left a record of *subsistence* activities, how they made a living. Nothing in the archaeology indicates a need to use complex language with metaphors or subjunctive tense. There were no sets of ornaments or figurines or paintings on the shelter's back wall. Anatomically modern humans lived for thousands of years in Africa without these luxuries, but when they moved into Europe 40,000 years ago, they developed societies demanding aesthetic enrichment. Hence, Upper Paleolithic sites of anatomically modern humans appear intellectually rich compared with Middle Paleolithic ones. Even Châtelperronian does not contradict this separation, allowing the inference that Neandertal brains had not evolved the capacity for art, nor perhaps for complex language. This doesn't mean the modern humans carried out a holocaust against poor, stupid Neandertals or that no intercourse took place. It may mean that over several thousand years, Neandertals became more and more marginal, failing to compete for the best resources against *sapiens*, finally becoming so reduced in numbers that their little communities did not reproduce their kind. Or it may be that anatomically modern humans moving out of Africa carried diseases for which they had developed immunity, but which devastated Neandertals. The interesting thing is that neither biological anthropologists nor archaeologists can fully and definitively describe Neandertal capacities, nor account for the extinction of the species.

■ ■ ■ SPREAD OF FARMING, SPREAD OF INDO-EUROPEANS

Relating human populations to cultures and languages looks obvious, but is full of controversy. Regarding Neandertals, their concern to protect loved ones' corpses from desecration by burying them bespeaks languages capable of expressing abstract ideas, that the dead body is not a thing to be abandoned but someone, special and well remembered. Anatomically modern Upper Paleolithic people painting strikingly lifelike animals deep inside dark caverns surely could speak poetically. Differences in artifact styles over time and between regions may reflect ethnic distinctions; part of the controversy over Clovis in America (Chapter 8) involves the issue of whether the continent-wide appearance of the technique and style is testimony to the rapid expansion of a parent population, like the spread of English in North America after 1607, or whether Clovis was imitated by non-related peoples—a person doesn't need words to learn flint-knapping from an expert.

More than a century ago, the German-American anthropologist Franz Boas attacked the standard notion that speakers of a language constituted a race. It looks pretty ridiculous to us now, the notion that there was an Anglo-Saxon race of English speakers born to dominate the world. English is a mixture of a host of other languages: Angle and Saxon, and other Germanic languages, with medieval Norman French added when Duke William of Normandy conquered England's king Harold the Saxon in 1066. Just to mix things up further, Duke William was Scandinavian by descent from Rollo who had conquered northwest France (Norman = Northmen) only 155 years earlier. Another way to see how untenable it is to tie "race" to language is to consider multilingual people: did Franz Boas become, in one person, multiracial when he added English to his native German, and then Inuktitut when he lived for a year in the Canadian Arctic with Inuit?

Contrary to the commonsense observation that whatever their biological parentage, humans learn their native language, or languages, from their caregivers, again and again the proposition crops up among archaeologists that the spread of an invention was linked to the spread of a population and its language. There are, indeed, some cases where this, roughly, happened, particularly the spread of Polynesian culture and Polynesian languages with the movements of populations of Polynesian physical types, entering central and eastern Pacific islands never before settled by humans, only one or two millennia before Europeans met and described these populations. It would be better to think of Polynesian colonization as an exception to Boas's rule that race, language, and culture don't usually link, rather than use Polynesians as model for the linked spread of cultural traits, language, and population. The usual situation, for many thousands of years, was for migrating people to displace or settle among indigenous populations, absorbing some of their language and culture, as happened with post-Columbian European invasions of the Americas. Notice, too, that although America's invaders imposed English, Spanish, French, or Portuguese upon the territories they conquered, the invaders themselves came from several regions of their home nations (for example, Scots, Welsh, and Irish as well as English came into North America), and in much of the Americas, soon brought in Africans.

Probably because the majority of archaeologists speak Indo-European languages, come from European populations, and descend from farmer ancestors, the spread of Indo-European languages and the spread of agriculture have been hypothesized to have occurred along with the expansion of populations that founded European nations. "Indo-European" is a large set of languages that share basic grammar and many words—for example, the distinction between nouns and verbs, and the word for "father" ("Vater," "pater," "padre"). The earliest written European languages, Greek and Latin, are Indo-European. Settled agricultural villages seem to have appeared earlier in southwestern Asia than anywhere else, and this region is in the middle of the range of Indo-European languages, which extends from India and Iran westward to Atlantic Europe. Therefore, agriculture as we know it, and Indo-European, could have spread together, as farmers in Turkey and the Near East sought more arable land. They would have halted in the Balkans, in southeast Europe, until they adapted their Mediterranean-zone crops and artifacts to survive in a temperate continental climate. Their spread eastward into Asia was halted by increasingly dry steppe conditions. Other complexes of agricultural crops, techniques, words, and artifacts developed elsewhere under different climates with their own indigenous plants —for example, rice and Asian millet farming in China, taro and yams in southeastern Asia and the Pacific, sorghum and African millets in Africa, and maize (corn) in America.

The maps drawn in textbooks greatly oversimplify the distribution of languages. When you get down to the details, maps can seem confusing. Smack in the middle of that nice, broad range of Indo-European languages—extending from England to India—are the totally unrelated Semitic languages (Figure 9.1). If European-type agriculture of wheat, sheep, and goats originated in the zone from Turkey to Iran, how do we explain the role of the people who spoke Semitic languages? They are indigenous to the eastern Mediterranean from Turkey to Egypt and as far east as Iraq, and are known to have been there for at least five thousand years from texts found in Mesopotamia. What role did they play in the origin of wheat agriculture? The case of Turkey, and its language Turkish, might provide an answer. Its present language, unrelated to either Indo-European or Semitic, has been used in that country for only five hundred years and was imposed on earlier populations by the conquests of Ottoman Turks from the central Asian steppes, who took the capital city Constantinople (Istanbul) in 1453. This historic scenario inspires an alternate theory on Indo-European expansion: that this language group, like the historic Turks, originated in western Asia in steppe country and spread by military conquests of indigenous agricultural peoples.

Detailed data, developed from a series of local archaeological records that can be strung together, show that apparently there were migrations of farmers into prime arable lands, especially along river valleys. Native hunter-gatherer peoples were sometimes quite willing to produce surpluses of those of their traditional goods that farmers wanted in trade—for example, fish or deer hides—rather than fight the newcomers. Analyses of bones from early farming villages in Europe indicate that in some cases, women from other localities moved into farming communities, presumably by coming as brides to their husbands' villages; possibly the women were from indigenous groups (as in the Americas, European invading men often took indigenous women as wives). Over time, distinct hunter-gatherer communities faded out, perhaps by becoming economically integrated with the farming villages.

We need to be careful, too, about what we mean by "agriculture." Most societies we know cultivated some plants and kept dogs, even if their economy

FIGURE 9.1 Map of Indo-European and Semitic languages. The white area in southeastern Europe is Hungary, speaking the Finno-Ugric language Magyar. Turkish now separates the Semitic block from Indo-European.

depended upon hunting, fishing, and harvesting wild plants. Historic hunter-gatherers populate regions where farming is not possible, such as the far north or deserts. Whether their limited cultivation, where it can be carried out, reflects contacts with agricultural nations, or was developed independently millennia ago, is difficult to tell—every ethnographically recorded population, including those described by the Classical Greek writer Herodotus two thousand years ago, has lived in a world of agricultural societies. Researchers distinguish between "agricultural societies" ("true agriculture") whose economy depends upon planting and cultivating selected food species, and others who may plant and cultivate foods but can survive without those particular plants. The evolution of agriculture began with hunter-gatherers managing favored foods, animal as well as plant, by removing competing organisms (weeding), diverting water where needed, and assisting in reproduction—for example, by sowing seeds.

In many localities, including the eastern half of North America, people were concerned to maintain a diversified resource base rather than clear out acres to raise one or a few crops that might fail from drought, hail, or locusts, leaving them in famine. Europeans invading eastern North America disdained those multi-crop fields and the practice of maintaining deer in adjacent woods as a livestock requiring little labor, the Europeans forgetting how often Eurasian agricultural economies reduced peasants to starvation. True agricultural economies did develop in the American Midwest and Southeast, sustaining relatively densely populated kingdoms, as on a larger scale agriculture developed earlier in Mesoamerica and tropical South America, feeding empires. The point is that agriculture is one economic strategy, and diversified food resources is another, quite intelligent one.

Sensible as it seems to avoid putting all one's eggs in one basket, diversity with reliance on indigenous plants has the drawback that it doesn't produce enough food to maintain large armies. Hunter-gatherer/minor-cultivator societies tend to favor spacing children several years apart so that mothers are not heavily burdened with frequent pregnancies, managing toddlers while nursing babies, and working to gather and process foods withal. Larger societies need more labor to produce farm surpluses to feed craftworkers, merchants, bureaucrats, and armies. Peasants stay near their plots, often by order of authorities, and women may leave infants in the village in order to work in the fields. This means the infants cannot nurse frequently, as they will if carried by the mother or cradled near her. Only frequent nursing inhibits a woman from becoming fertile again; so peasant women who don't have their babies with them all the time will become pregnant again sooner than will a woman whose baby stays in a sling close to her breast. Agricultural societies don't expand because they produce a lot of food, they produce a lot of food because farmers are forced to feed many demanding non-farmers. To ensure there will be enough farmland and farmers to support them, the ruling class looks to conquer more lands, for which they need plenty of soldiers, for which they need more land.

Archaeology indicates that agriculture spread about three thousand years before the rise of conquest-minded states supporting armies. Most of those states were literate, so we know what languages they used—and thus we know that some states were multilingual. Historic distributions of languages owe much to state expansions, the way English spread over North America with armed force behind it, taking over areas speaking Algonquin (New England), and later, French (Louisiana), Spanish (New Mexico), and Russian (Alaska, Northern California). Unlike the historical case of North America, hypothesizing that the spread of Indo-European was related to the Neolithic spread of agricultural villages beginning about 7000 B.C.E. comes up against the complete lack of direct data on languages before literacy. With the Bronze Age, three thousand years later, we do have texts (in Greece, Egypt, and western Asia), and we have cities and armies. But we don't have an explanation for the Semitic languages and peoples, surrounded by Indo-Europeans. Nor are there words in common among Indo-European languages for things like wheat agriculture, sheep and goats, and settled villages. We would expect this if a single Indo-European agricultural population spread across the region. To the contrary, linguistic analyses of "Proto-Indo-European" indicate its speakers used horses and horse-drawn wagons, were warlike, and were familiar with plants growing north of Turkey, as well as wheat. A better inference from our data is that farming practices, the development of languages and dialects of Indo-European, and movements of people resulted from combinations of migrating communities, conquering armies, and trade between Indo-European speakers and neighboring peoples. Boas was correct in insisting that biology, language, and culture are not tied together.

■ ■ ■ WAR AND CANNIBALISM

Archaeologists can use the abundance of historic data—as much as five thousand years of it in Mesopotamia—to look at whether language spreads might have resulted more often from conquests than from a peaceful Neolithic farming expansion. On the one hand, picturing those early farmers armed to overcome hunting-gathering natives may be a more realistic scenario than simply seeing the farmers moving along river valleys while the natives fade into the hills or bestir themselves to bring fish and game to the newcomers, like kindergarten pictures of happy Pilgrims and Indians at the Plymouth Thanksgiving. On the other hand, historical data tell us, above all, that invasions and wars imposing conquerors' languages do not produce long-term stable cultures; another conqueror rides over the horizon and repeats the cycle. Research into place names uncovers palimpsests of ancient cultural succession, as in England where Roman, Anglo-Saxon, Danish, and Norman French names on a map attest to those successive invasions, or in America where First Nations' place names coexist with English, French, or Spanish. Acknowledging the antiquity and pervasiveness of warfare leads to the inference that attributing distributions of languages to migrations of early farmers is a weak explanation.

Back in the seventeenth century, the English philosopher Thomas Hobbes reasoned that early humans endured miserable lives—"solitary, poor, nasty, brutish, and short," he said—constantly fighting "where every man is enemy to every man." He supposed that people agreed to live civilly when agriculture developed, in order to protect their farms and goods. Many writers pictured a "wished-for past" of peaceful villagers devoted to their cultivated plots, ended by onslaughts of fierce nomadic warriors (see Marija Gimbutas's projection of Goddess-worshipping Neolithic villagers in southeastern Europe, Chapter 6). American archaeologists generally accepted the idea that you wouldn't have war unless you had states with armies; smaller societies might have feuds or raids, but overall, when population density was relatively low, people presumably had little incentive to conquer more land. This "paradigm" (model) of a peaceful Neolithic was so taken for granted that archaeologists ignored the evidence staring them in the face, such as excavated skeletons showing evidence of violence. That was especially true in the American Southwest, where the Pueblos were viewed as idyllic communities treasuring harmony above all virtues. Biological anthropologist Christy Turner aroused virulent opposition when he insisted on publishing, in 1999, the results of his, and his wife's, years of examination of thousands of human bones, of which a small but significant number bore marks of lethal trauma and even butchering.

The power of stereotype can be seen in the persistence of the image of "peaceful Pueblos," apparently a stereotype coming from Pueblo Indians' unwillingness to fight the United States in battles. By the time the United States wrested the Southwest and California from Mexico, in 1846, the Pueblos had learned that Euroamericans' numbers and military technology made open warfare against them a poor strategy. This was not always true. The 1680 revolt of an alliance of Pueblos against Spanish conquerors succeeded for twelve years. Then the Spanish came back with larger forces, reconquered the Southwest, and subjugated the Pueblo peoples again. Two centuries later, Pueblos superficially cooperated with American occupying forces while moving their cultures underground, literally so in the sunken kivas built inside pueblos and used for rituals and meetings. Pueblos did value harmony highly, to the point of enforcing it by executing members suspected of witchcraft or subversion. There was plenty of evidence that Pueblos had soldiers' organizations, deities to favor them in war, legendary histories of wars, and claimed annihilations of enemies. The famous cliff dwellings in Mesa Verde, Colorado, bespeak communities so desperate for protection against enemies that they lived in nearly inaccessible clefts in cliffs (Figure 9.2). Tourists coming to see these beautiful, "mysterious" canyon sites were told of "a vanished people" rather than informed, as the National Park Service now does, of warfare. Refugees eventually left the Mesa Verde area and joined more secure pueblos in the Río Grande Valley.

The myth of peaceful Pueblos began to change in the 1990s. There doesn't seem to have been any one revolutionary study that sparked it, although Turner's

FIGURE 9.2 A Mesa Verde pueblo, Colorado, ca. 1250. *Photo by Alice B. Kehoe.*

insistence on the abundance of clear empirical evidence of violence was a factor. A co-authored paper by two well-respected Southwestern archaeologists in a volume called *Themes in Southwest Prehistory* (Wilcox and Haas 1994), and then publication of a book explicitly focusing on archaeological data for *War Before Civilization* (Keeley 1996) further raised the issue. The latter book attacked the notions of both peaceful Neolithic and peaceful Pueblos, showing instead that all over the world, settlements were often situated defensively and fortified, and their inhabitants often brutally killed. Sites contained contorted, hacked, or bashed skeletons, left unburied or thrown into ditches or rooms, and then abandoned. In some gruesome cases, dozens or even hundreds of women, babies, and children as well as men, were strewn on the ground or piled into a heap, their arms, shoulders, and heads broken. The evidence could only be interpreted as massacres. The final taboo, cannibalism, had to be recognized when unequivocal data were produced. In addition to Turner's butchering cuts on human bones, and meatier human bones found at cooking hearths, chemical analysis of dried human feces near one such hearth demonstrated human myoglobin that must have come from a person eating and digesting human flesh.

Accepting evidence for "war before civilization" (its author, Keeley, meant "before the development of states") leaves unsettled the question of *why?* Mainstream American archaeologists wanting a rational explanation seek data showing that climate change or population growth has had an impact on food supplies, leading to shortages that resulted in conflict. Correlations between climate, food shortage, and violence can be adduced, but we must always be wary of thinking that correlations give us cause and effect. A classic example of mistaking correlation with cause and effect took place during World War I. U.S. Secretary of State William Jennings Bryan became heartsick over the terrible rampages of the war, even though it did not yet involve American troops. He read a book attributing war to a breakdown of morals stemming from the secular mind-set associated with Charles Darwin's thesis of evolutionary change. Equating belief in evolution with the cause of war, Bryan began to vociferously denounce "evolution." World War I, the cruel end to the late nineteenth century's belief in social progress toward prosperity and peace, obviously did not result from people accepting the concept of "descent with modification" (Darwin's actual term) to explain change in organisms through time—there were countless wars before Darwin! An archaeologist hoping to find a straightforward cause for a phenomenon may seize upon data fitting a hypothesis, neglecting the good scientist's duty to weigh alternate hypotheses. Climate change and population growth stressing societies' capacity to support their people certainly did, in some instances, spur them to war upon neighbors, but neither seems to have been a *necessary* condition for warfare.

■ ■ ■ CASE IN POINT: MOUND 72, CAHOKIA

From about A.D. 1050 to 1275, a state with its capital city at present-day East St. Louis (Illinois) dominated the Midwest. Its location along the rich floodplain at the confluence of the Mississippi and Missouri Rivers made it the hub of trade and the arts, like its modern successor St. Louis, Missouri. Its huge mounds and grand plazas are still awesome, worthy of their place on UNESCO's list of World Heritage Sites. Because the site's people left no writing, we don't know the native name of the ancient city we call Cahokia, nor the titles or names of its leaders, nor what propelled them to build so impressive a capital. During its first century, Cahokia had no apparent defenses other than its sheer size and control points at narrows or on bluffs along the rivers approaching it. Then, about 1150, Cahokians built a tall timber palisade three kilometers (two miles) in circumference around its "downtown" zone. At this time, subsidiary upland villages and farmsteads were abandoned, and at least one compound with an upper-class residence and a maize granary was burned. How many may have been torched, we can't tell because so little excavation has been carried out in East St. Louis, where surveys indicate almost continuous Cahokian residences and farms. A century later, Cahokia itself was almost a ghost city, and political power passed to lesser kingdoms in the South, and to independent communities in the northern Midwest.

Early excavators of mounds in Cahokia encountered human skeletons, elegant Ramey Incised pottery, finely flaked oval knife blades, and thousands of beads and pendants made from shells imported over a thousand miles from the Gulf of Mexico or Florida. Crude archaeological methods of the time, employing teams of men with shovels and horse-pulled draglines, destroyed much of the information we crave about the contexts of burials and valued artifacts. Archaeology later in the twentieth century revealed that the great mounds, up to one hundred feet (thirty meters) high, cover stages of building and series of colored clay overlays, suggesting the mounds we see represent multi-event rituals, and originally carried veneers of black, white, and blue clays. The scale of Cahokia and its major mounds frustrates archaeologists, for all we can do is sample, never seeing the original occupation surface wholly exposed. Nor would we today destroy a monumental mound just to see all it might hold. That would be robbing future generations.

One archaeologist decided in 1967 that a doable project for his annual field school would be to select a small mound outside the Grand Plaza area for excavation. This being the time that archaeoastronomy was taking off, he chose a little mound, #72 on the map of Cahokia, in a direct line south from one side of the central greatest mound, but beyond the big mounds at the south end of the Grand Plaza (Figure 9.3). Mound 72 turned out to astound everyone, for it contained 272 bodies, of which 266 may have been sacrifices in a ritual comparable only to legendary Aztec sacrifices. The other half dozen were much later graves probably from natural deaths. Nothing like this enormous sacrifice has been seen anywhere else in the United States, from any time period.

Initially, the excavator thought he had a family tomb like the stone-built "barrows" of England, where over a century or more, the tomb was opened and reopened for each death in the extended family. When the entire Mound 72,

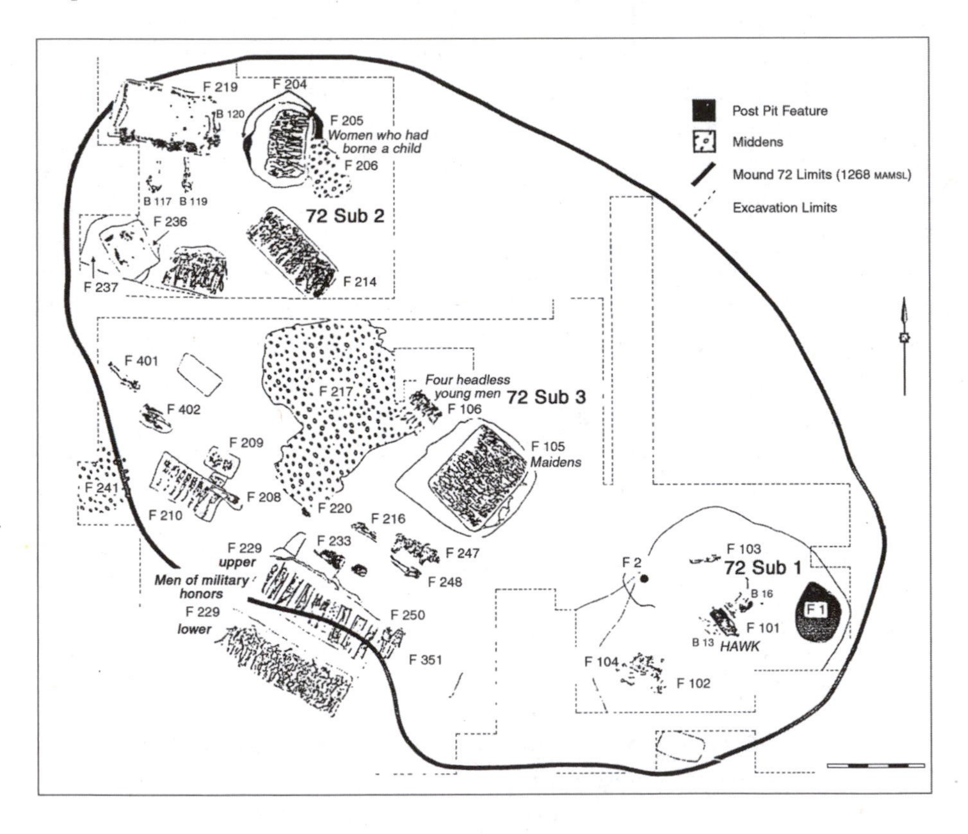

FIGURE 9.3 Mound 72, Cahokia (Illinois), ca. A.D. 1100, showing burial pits with sacrificed people. *Modified by Alice B. Kehoe after figure 1.6 in* The Mound 72 Area, *Illinois State Museum Reports of Investigations No. 54. Used by permission.*

45 meters by 15 meters, had been excavated, it became clear that it covered a single prolonged event, with dozens of people killed in groups and laid in pits, one after another from southeast to northwest along the long axis of the mound base. Highly valued artifacts were sacrificed, too, laid in other pits near the sets of bodies. A biological anthropologist was brought in to examine the skeletons *in situ* (in place) to obtain maximum information about the extraordinary find; it was particularly important to have this specialist there because preservation was poor, making it difficult to lift bones out without their crumbling. Chemical analyses of some of the bones revealed that those people had been brought to Cahokia from other regions, as had the mica sheets, copper, and beautiful stone arrowpoints in the artifact offerings. If ever anyone doubted the power of the Cahokian state, that scene of hundreds of victims, women and men, led to slaughter before the citizens and visiting dignitaries would testify to Cahokia's supremacy.

Mound 72 dates to the late eleventh century A.D., the climax of construction at Cahokia but probably decades before the palisade was built. This was not a massacre by enemies; the victims were sorted and their burials orderly; besides, the valuable offerings would never have been given up by hostile soldiers. Unlike many sacrifice rituals discovered in Mexico, the unfortunates in Mound 72 were not offerings for divine favor upon a new building, for there was none on the ridge-shaped mound. Mound 72's distance from the Grand Plaza, its being out of sight behind the Grand Plaza's great perimeter mounds, suggests the sacrifice ritual was not part of the state's rituals of governance performed on the Grand Plaza. Puzzling over this anomaly, I noticed that the series of sets of victims parallels a chant in the Osage nation's ritual for success in war:

> They [Osage warriors] gave a forward stroke . . .
> It is a youth in his adolescence . . .
> They gave a second stroke . . .
> It is a maiden in her adolescence . . .
> They gave a third stroke . . .
> It is the man who is honored for his military achievements . . .
> They gave a fourth stroke . . .
> It is the woman who has given birth to her first child . . .
> On going against the enemy
> We shall always overcome them with ease, as we travel the path of life, they said to one another. (La Flesche 1930:712–713)

This chant accompanied a medicine bundle (wrapped set of holy objects) consisting of a rectangular woven mat with the lower end turned up to make a pocket holding a falcon hawk representing the Osage soldier's swift and lethal strikes (Figure 9.4).

FIGURE 9.4 Shrine of the Wa-Xo'-Be (Hawk), the war medicine bundle of the Osage, showing rolled mat, enclosing deerskin and woven bison-hair bags, and final wrapping with eagle claw and scalp. *Francis La Flesche, Smithsonian Institution B.A.E. 45th Annual Report, Pl. 25.*

Mound 72 has at its southeast end a man placed on a blanket in the shape of a hawk with folded wings (like the dead hawk in the medicine bundle), entirely covered with shell beads. Another man lies underneath as if bearing the hawk. The heads of both the principal man and the hawk blanket beneath him lie to the southeast. If the mat for the Osage war bundle is unrolled flat, the hawk lies in the south end, just like the hawk-man in Mound 72. Closest to him under the mound are young men, then a pit with fifty-three maidens (young women whose skeletons indicated they had never borne children), then a pit with two layers of bodies. On the top there was a row of bodies on litters such as were historically used to carry aristocrats in Southeastern First Nations. Below the litters rested the remains of the litter bearers, brutally killed and thrown into the pit. The persons on the litters could have been enemy men accorded military honors. Northwest of these are three additional pits with nearly two dozen women in each, including women who had borne children. Preservation of bones in the mound was not good, so not all female skeletons could be definitively analyzed for this. In other words, Mound 72 visualizes the Osage war chant, first the young men, then the maidens, then honored men, and then mothers.

Skeptics will say that nineteenth-century Osage priests could not have retained ritual from seven hundred years earlier. But their holy texts were believed to be innately powerful, requiring very careful recitation by well-trained, ordained priests; the Omaha Indian ethnographer, Francis La Flesche, who recorded the texts, was warned that the words could kill the uninitiated. Maybe words believed so powerful could have been passed down carefully over seven centuries. At European contact in the late seventeenth century, the Cahokia area was virtually a no-man's-land frontier between the Osage, occupying resource-rich territory up the Missouri River, and other First Nations on the Illinois side of the Missouri-Mississippi confluence. The Osage were the most powerful Midwest nation then and in the eighteenth century, and fought the United States in the nineteenth century until the post–Civil War final campaigns against First Nations. Most ethnohistorians consider it reasonable that the Osage and related Omaha may be descendants of Cahokia.

Other priestly Osage texts contain passages possibly reflecting Cahokia's rituals, including one set describing thunderclouds moving in to ring the city like high hills—were the great mounds around the plazas meant to be thunderclouds harboring thunderbirds?

> "Behold the beauty of yonder moving black sky.
> "Behold the beauty of yonder moving gray sky.
> "Behold the beauty of yonder moving white sky.
> "Behold the beauty of yonder moving blue sky."

> In this dramatic fashion the ancient Non'-hon-zhin-ga [Osage priests] have given expression to their conception of the inseparable unity of the Sky and the Earth out of whose combined mystic power the great pageant of life goes forth on its endless journey. (La Flesche 1925:349–355, 360)

In a plaza west of Cahokia's Grand Plaza, there was an arc of very large cedar posts, probably part of a circle (the excavation project couldn't follow the whole arc), with a single big post just offset from the center. The excavator, in the 1960s, thought this might be a "woodhenge," a First Nation version of Stonehenge, for sighting the solstice sunrise over the largest mound to the east (see Chapter 8, on archaeoastronomy). This could have been done, but perhaps the circle represented the "forest of cedars upon which thunderbirds alight" in the Omaha religious text describing the finding of the magical tree embodying the spirit of their nation. The huge post in the center of the circle then would be a Cahokia-scale version of the pole that historic Omaha kept in its own tipi, their Venerable Man in his shrine (see Chapter 4).

I can't prove a link, nor that there wasn't a link, and not all my colleagues agree with my interpretation. Equally controversial, and equally unproven, are the similarities between Cahokia symbols, Osage priestly texts, and Mixtec texts. These writings, from a contemporary people living in central Mexico, indicate contacts between Mexico and Cahokia, and are embedded in Osage traditional history of their origin. Mound 72's astounding, and appalling, series of human sacrifices raises the possibility that Cahokia's rulers wanted to imitate their contemporaries in Mexico, in ritual as well as in urban plan. The layout of the sets of persons, paralleling the Osage war chant, and the resemblance of the "woodhenge" to the closely related Omahas' image of thunderbird roosts around a mighty tree, could link Cahokia to these nations dominating the Midwest four centuries later when Europeans arrived. And if we can plausibly link this specific site, Cahokia, to likely descendant nations, Osage/Omaha, we can infer the Cahokians spoke a Chiwere Siouan language like Osage and Omaha, ritualized the practice of aggressive war as they did, conceptualized a forest home of thunderbirds surrounding a tree of life, and "the great pageant of life" within "the inseparable unity of the Sky and the Earth," tied by great mounds like hills. I see many data that support the interpretation of a strong cultural link between Cahokia and the historic Osage.

Controversies abound over Cahokia. Mainstream American archaeologists following the nineteenth-century social-evolution model have resisted recognizing Cahokia as a state, terming it instead a "chiefdom" (see Chapter 10). One of these archaeologists calculated that a total population of only eight thousand men, women, and children, spread in little farms over the river plain, could have built the big mounds basketload by basketload during the off-season for agriculture. *Why* simple farmers in the heart of the flat Midwest would build artificial mountains, he didn't say.

Another long-running controversy concerns the possible ties between Cahokia and Mexico. Cahokia's city center is dissimilar to any other Anglo-American site, but its grid of mounds around rectangular plazas is similar to Mexico's prehistoric cities. Little other than the architecture resembles Mexican cultures, but there were a few burials near Cahokia of people who had been beautified by having their front teeth filed to points. This was common fashion in Mexico at the time of Cahokia, but not practiced elsewhere in Anglo-America except for a few at a contemporary unusually large town, in Chaco Canyon in northern New Mexico, a thousand miles west and ancestral to Pueblos.

During the same period that Cahokia dominated the Midwest, Chaco dominated the Southwest. Its architecture does not follow the Mesoamerican mounds-around-plazas blueprint used in Cahokia; Chaco's people built defensible multistory pueblos with blank outer walls and heavily guarded entrances. At Chaco, however, there is direct evidence of trade with central Mexico: turquoise mined by Chacoans has been found in Mexico (it can be sourced chemically to New Mexico), and astonishingly, bones and eggshells from scarlet macaw parrots native to southeastern Mexican tropics lie in stone pens in Chaco. These breeding birds had been carried in porters' backpacks (pictured on painted pottery) thousands of miles to the dry Southwest. Still today, Pueblo priests use scarlet macaw feathers (mostly donated by pet shops and zoos) in ceremonies. The American Southwest's contacts with Mexico over more than two millennia, bringing in maize agriculture, pottery-making, house styles, and religious symbols and concepts, were generally disregarded by previous generations of American archaeologists. Over time, the weight of evidence has shifted the paradigm, so that most Southwestern archaeologists now recognize regularized contact, though they still disagree about the significance of Mexican influence.

While the biological and chemical evidence—scarlet macaw remains and turquoise from New Mexico mines—for Mexican–American Southwest contacts has overcome reluctance to recognize Mexican links, such strong evidence has not, so far at least, been found at Cahokia. Cahokians apparently did not import and breed live macaws, and they could not export precious stones, since none occur in their domain. If Cahokia did trade with Mexico, it could have exported tanned deer hides, dried meat, maize, and/or slaves, all of these abundant in the Midwest but in short supply and high demand in Mexico. These four products were commercially produced and traded among First Nations before European invasions, and were primary trade goods with Europeans afterward.

We know about the historic markets from European documents, but the very few pre-European Mexican documents that survive do not refer to kingdoms north of the Gulf of Mexico, so far as has been deciphered. It seems likely that the four principal products from contact-period First Nations named above were extensively traded between North American nations, and to a much lesser extent traded across the Gulf of Mexico to pre-contact Mexican kingdoms in exchange for fine garments, brilliant feather headdresses and cloaks, and, as in the Southwest, priestly knowledge. The lords of Cahokia probably traded in this way with the rulers of Mexico's Toltec empire, the dates of which coincide remarkably closely to dates for both Cahokia and Chaco. Perhaps the Toltecs, themselves ambitious to surpass all others, encouraged these distant entrepreneurs to think big, to build up long-distance trade, and in the case of Cahokia, to copy the urban plan attributed to fabled Tollan, an idealized prototype of Mexico's indigenous cities. Unfortunately, all these probable goods were perishable, and the evidence for this trade is still slim.

Overall, collaborations between biological anthropologists and archaeologists have confirmed evidence for warfare and, at Mound 72, human sacrifice, while otherwise tending to discourage simplistic assumptions that material culture equates to populations, "pots equal people." Genetic and other biological traits are not wired to either language or culture, not even with Neandertals. This chapter has focused on controversies where biological anthropologists have been able to advise archaeologists. The bulk of collaborations between the two fields has been in identifying sex, age, cause of death, pathologies, and life experience (nutrition, illness episodes, physical labor) in human remains recovered by archaeologists. These data provided by biological anthropologists build pictures of communities, complementing archaeologists' information on the material culture of those communities. Most of the work, happily, does not involve controversies.

SOURCES

Arkush, Elizabeth N., and Mark W. Allen, editors, 2006. *The Archaeology of Warfare: Prehistories of Raiding and Conquest.* Gainesville: University of Florida Press.

Bellwood, Peter, 2005. *First Farmers: The Origins of Agricultural Societies.* Malden, MA: Blackwell.

Buikstra, Jane E., and Lane A. Beck, editors, 2006. *Bioarchaeology: The Contextual Analysis of Human Remains.* Amsterdam: Academic Press/Elsevier.

Fowler, Melvin L., Jerome Rose, Barbara Vander Leest, and Steven R. Ahler, 1999. *The Mound 72 Area: Dedicated and Sacred Space in Early Cahokia.* Reports of Investigations, No. 54. Springfield, IL: Illinois State Museum.

Fox, Richard G., editor, 1998. The Neanderthal Problem and the Evolution of Human Behavior. Special issue of *Current Anthropology* 39, Supplement.

Haas, Jonathan, and Winifred Creamer, 1997. Warfare among the Pueblos: Myth, History, and Ethnography. *Ethnohistory* 44 (2):235–261.

Harris, David R., editor, 1996. *The Origins and Spread of Agriculture and Pastoralism in Eurasia*. Washington, DC: Smithsonian Institution Press.

Henry, Donald O., Harold J. Hietala, Arlene M. Rosen, Yuri E. Demidenko, Vitaliy I. Usik, and Teresa L. Armagan, 2004. Human Behavioral Organization in the Middle Paleolithic: Were Neanderthals Different? *American Anthropologist* 106(1):17–31.

Hobbes, Thomas, 1651. *Leviathan, or the Matter, Form and Power of a Commonwealth, Ecclesiastical and Civil.* First published in London.

Keeley, Lawrence H., 1996. *War before Civilization: The Myth of the Peaceful Savage.* New York: Oxford University Press.

Kehoe, Alice Beck, 2006. *North American Indians: A Comprehensive Account*, 3rd edition. Upper Saddle River, NJ: Prentice-Hall.

—. 2007. Osage Texts and Cahokia Data. In *Ancient Objects and Sacred Realms*, edited by F. Kent Reilly III and James Garber. Austin: University of Texas Press.

La Flesche, Francis, 1925. The Osage Tribe: Rite of Vigil. 39th *Annual Report*, Bureau of American Ethnology, Smithsonian Institution, pp. 31–635. Washington, DC: Government Printing Office.

—. 1930. The Osage Tribe: Rite of the Wa-Xo'-Be. 45th *Annual Report*, Bureau of American Ethnology, Smithsonian Institution, pp. 528–833. Washington, DC: Government Printing Office.

LeBlanc, Steven A., 1999. *Prehistoric Warfare in the American Southwest.* Salt Lake City: University of Utah Press.

Pauketat, Timothy R., 2004. *Ancient Cahokia and the Mississippians.* Cambridge: Cambridge University Press.

—. 2008. *Cahokia's Big Bang and the Story of Ancient America.* New York: Viking/ Penguin.

Pettitt, Paul, 2005. The Rise of Modern Humans. In *The Human Past: World Prehistory and the Development of Human Societies*, edited by Chris Scarre, pp. 124–173. London: Thames and Hudson.

Rose, Jerome C., 1999. Mortuary Data and Analysis. In *The Mound 72 Area: Dedicated and Sacred Space in Early Cahokia*, by Melvin L. Fowler, Jerome Rose, Barbara Vander Leest, and Steven R. Ahler, pp. 63–82. Reports of Investigations, No. 54. Springfield: Illinois State Museum.

Wilcox, David R., and Jonathan Haas, 1994. The Scream of the Butterfly: Competition and Conflict in the Prehistoric Southwest. In *Themes in Southwest Prehistory*, edited by George J. Gumerman, pp. 211–238. Santa Fe, NM: School of American Research Press.

10 Competing Theories of Cultural Development

How do we distinguish humans in the past from other animals? We look for evidence of culture: surviving through extra-somatic ("outside the body") means is necessary for our species. Other species use teeth to render food digestible; we use knives and pestles and milling stones. Other species grow from their bodies the insulation and protective shells they need; we make clothing and armor and shelters. No other species is so completely dependent on extra-somatic aids. No other species dares to manipulate fire. Archaeology is the science that can tell the history of humans' remarkable inventions, culminating in today's billions of people nourished by agriculture, living in towns and cities, instant-messaging around the world, saved from death by disease or injury—or killed by weapons of nightmare firepower.

Most archaeologists deal only with sites occupied by anatomically modern humans near the end of, and after, the Pleistocene Ice Age. The deep question when we get to this last stage of our physical evolution is why, and how, cultures change. With archaeological data covering the entire globe and all of human history, archaeological theorists have attempted to answer this overarching question. One extreme position in the debate is that cultures are superorganisms—groups, like people, that are constantly growing, maturing, and dying. At the opposite pole, theorists contend that human societies constantly try to organize diversity but cannot inhibit it for any length of time, and therefore cultures are unstable. "Superorganic" was a nineteenth-century idea about a "vital force" weak in "primitive societies" and vigorous in civilized ones, particularly the writer's own social class and country. To accommodate history's chronicle of changes in nations' political and economic fortunes, proponents of the superorganic idea wrote about civilizations being born, growing, maturing, declining, and dying. Classical Greece or the Ottoman Empire could be described in this manner as superorganisms. The burning issue was whether enlightened social scientists could save our own civilization, or our own nation, from inevitable decline and death. Most believers in superorganic cultures feared we could not.

By the mid-twentieth century, the idea of literally superorganic cultures came to be recognized as metaphysical, that is, beyond the means of science to investigate. The idea was politely labeled a metaphor—a society's fortunes may grow and decline *as if* it were a living creature but it really is not.

This chapter examines controversies over theories proposing to explain cultural developments, including theoretical positions labeled evolutionary archaeology, processual and postprocessual archaeologies, "agency," political archaeology, and realpolitik. I'll explain these jargon terms soon. They all relate to deep questions about the essence of culture and human nature.

■ ■ ■ CULTURAL EVOLUTION

While the concept of superorganic cultures was discarded, a related idea, that cultures *evolve*, persisted. Its formulation is also nineteenth-century but predates Charles Darwin. The same year, 1844, that Darwin first wrote up a draft of his thesis, Scottish journalist Robert Chambers published his "scandalous" *Vestiges of the Natural History of Creation* (see Chapter 6), outlining in it his belief that societies evolve by the occasional birth of geniuses who invent new tools, art, or political organizations. Chambers, a strong advocate for democracy, insisted that geniuses are born into all human populations ("races" in nineteenth-century terms). He attributed the greater complexity of some societies, such as his own Britain, to sharing within trading alliances with neighboring groups. Simple societies are those isolated from sharing networks by mountains, trackless forests, oceans, or deserts. Chambers's ideas were not generally accepted. His contemporaries preferred either a racist hierarchy attributed to God favoring Europeans, or unequal evolution, where some societies evolved very slowly or not at all while others were imbued with vital force that allowed them to evolve at a rapid pace. Darwin and another naturalist, Alfred Wallace, published in 1859 the evidence and explanation of descent with modification through natural selection for characteristics assisting survival and reproduction. Thinkers who rejected the concepts of vital force and superorganic cultures could more readily accept the thesis that social behavior might, like physical traits, be subject to natural selection. Probably everyone today agrees that natural selection for survival and reproduction can act upon cultural traits. An example could be the extraordinary distribution of Clovis blades (see Chapter 8), which some argue spread rapidly across the continent because they made more lethal spear-points, contributing to survival through improving hunting success.

Equating evolution with "Progress" bedevils discussion of cultural evolution. Linking the two was a nineteenth-century Western belief that their civilization was hurtling forward like a railway train (their symbol of modern technological triumph) toward material abundance, peace, and domestic bliss. Thomas Edison brought us electric lights and the phonograph, Samuel Morse and Alexander Graham Bell gave us telegraph and telephone communication, railroads spanned continents and steamboats the oceans in days not months, mass-produced goods were mass-merchandised in department stores (as exciting then as later shopping malls). Oh, the list of wonderful inventions goes on and on! Only, was it progress when cities were blanketed in industrial smog? When millions labored twelve-hour days tending whirring machinery or in tunnels hacking coal for factories? When beggars, drunks, and abandoned children haunted the streets? Charles Dickens, the novelist who as a boy had been forced to work in a shoe-polish factory, exposed the misery on the underbelly of English civilization.

Mixed up with the vision of Progress (capitalized by many writers) was the belief that evolution always, or normally, goes from simple to complex. See, the earliest organisms were one-celled creatures, and after eons of evolution we have complex multi-celled organisms such as ourselves, sharks, trees, and roses; likewise, artifacts evolved from simple hand-held stone tools to modern machinery, and societies from little bands to empires with millions of people. Nineteenth-century evolutionists went so far as to assert it is a law of nature that evolution proceeds in only one direction. Instances of societies going the other way—for example, the Roman Empire breaking up into small kingdoms—could be attributed to moral corruption or the fierce attacks of Attila the Hun and other nomadic hordes. Decline or degeneration didn't invalidate linking evolution with Progress because, after all, Western Europe's kingdoms did afterward evolve into complex modern nations, and the Huns and their ilk settled down to be absorbed into civilization.

Nineteenth-century cultural evolution was challenged by two men coming from secularized German Jewish families, Karl Marx and Franz Boas. Marx (1818–1883) and his close friend and supporter Friedrich Engels focused on the working class, the proletariat. Observing exploitation of the working poor, Marx researched history and concluded that conflict between social classes—rich masters and oppressed laborers—boils over from time to time in revolutions. He expected the proletariat would rise against capitalists and aristocrats, organize across national boundaries, and restore "primitive" communal management (socialism), promoting democracy and eliminating poverty. Needless to say, Marx was denounced.[1] Marx's view that stratified social classes foment conflict, revolution, and in time create a new political-economic order did not attract nineteenth-century archaeologists. It did intrigue an Australian-born British archaeologist, V. Gordon Childe, in the 1930s. Childe realized that Marx's emphasis on political economics could assist archaeologists in inferring social conditions from material culture. Marx's relegation of religion and ideology to the "superstructure" of society—the base being how resources were obtained and distributed—fit the limitations of the archaeological record, especially in prehistory, where religion and ideology left few and ambiguous traces. Childe's "Marxism" was pragmatic, not doctrinaire; he combined it with close attention to detailed data he collected visiting sites and museum collections throughout Europe, working out hypotheses of intersocietal contacts and movements. His has been a hard act to follow—he was totally dedicated to work, spoke a number of languages, and had a phenomenal memory for artifacts and site features. His theories of cultural development still have many adherents.

Franz Boas (1858–1942), forty years younger than Karl Marx, like him experienced the outsider status Germany imposed on its Jewish citizens. Boas emigrated to the United States after earning a Ph.D. and spending a year living with Inuit in Baffin Land. His Arctic experience, trying to pull his weight with his native hosts (literally, as they hauled boats or sleds over ice), taught him to respect human ingenuity in overcoming the most difficult environments. He was surprised that in that barren, colorless snow world, people joked, sang, composed poetry, and played games. The complete humanity of his fur-clad companions gave him a perspective on "primitive peoples" at odds with the theory of cultural evolution held by writers comfortable in their civilized libraries. Boas became an influential professor of anthropology at Columbia University in New York, teaching that all nations are the products of their particular histories, formed by intricate combinations of environment, intersocietal contacts, and their own thinkers and artists. He strongly and actively opposed racist doctrines, doing all he could to get higher education and academic appointments for gifted Africans, African-Americans, and American Indians, as well as students from poor immigrant families, and women. Archaeology was not among his own research activities, but he sponsored and fostered archaeological research projects, institutions, and practitioners. Impatient and irritated by sloppy reasoning

[1] A radical American woman, Victoria Claflin Woodhull (1838–1927), published Marx's *Communist Manifesto* in translation in her weekly newspaper and was jailed; see the "Notes" section at the end of the book.

and poorly supported interpretations, Boas denounced popular but racist notions of cultural evolutionism, in particular as expounded by the nineteenth-century American lawyer Lewis Henry Morgan, author of an elaborate scheme of cultural evolution in his book *Ancient Society*. Morgan claimed, for example, that descriptions of Aztec civilization were no more than conquistadors' lying boasts, that the Aztecs had never really had an empire.

If Morgan's blatant prejudice irritated Boas, Boas's scorn for the American lawyer-turned-pioneer anthropologist irritated a younger American, Leslie White (1900–1975). White espoused a radical socialist politics that he was forced to hide if he wished to retain his professorship at the University of Michigan during the Joseph McCarthy era of the 1950s, when many academics were fired for holding similar beliefs. White read Morgan's *Ancient Society* because Marx and Engels had admired it. Anti-Semitism was common in America in White's youth, as was disdain and fear of immigrants, so it isn't surprising that White disliked Boas. He was jealous of his influence, and wanted to champion the American pioneer anthropologist that Boas put down. Leslie White took up cultural evolutionism and recast it as a very American story of technological progress hinging on the harnessing of ever more energy. He made it look scientific by claiming that human history is summarized by a simple formula, $C = E \times T$ (Culture equals Energy times Technology, looking like Einstein's $E = MC^2$). Evolutionary progress, he wrote, comes through capturing first animal muscle power and then mechanical power. His books came out in 1949 and 1959, when the United States had become the world superpower by annihilating two Japanese cities with the incredible energy harnessed in atomic bombs.

Leslie White attracted veterans from World War II, men eager to build a better world than the havoc they had suffered through. There is a peculiar effect noted by the nineteenth-century social historian Henry Adams, a young man during the Civil War; as he put it,

> [u]nbroken Evolution under uniform conditions . . . Such a working system for the universe suited a young man who had just helped waste five or ten thousand million dollars and a million lives, more or less, to enforce unity and uniformity on people who objected to it; the idea was only too seductive in its perfection. (Adams 1918:226)

Although Leslie White was a theorist, not a field archaeologist, the postwar University of Michigan employed a couple of leading archaeologists whose students took anthropology courses from White and his colleague Elman Service, who also promoted a cultural evolution schema. Service taught a simple evolutionary progression of human societies from bands, to tribes, to chiefdoms, to archaic and then modern states. Ethnologists criticized his scheme for being too much an academic construct, not well grounded in a sufficient range of ethnographic data. Its simple, logical set of categories made it easy to use, to slot in data—"the idea was only too seductive in its perfection." The cohort of Michigan students excited by White's and Service's unilinear cultural evolution scheme went out in the 1960s to throw down Boas's historical particularism and replace it with their "New Archaeology."

NEW ARCHAEOLOGY AND ITS SUCCESSORS

New Archaeologists of the 1960s and 1970s insisted that they were true scientists methodically testing hypotheses (the hypothetico-deductive, or H-D, method; see Chapter 2). They configured data into closed-system diagrams that look like plumbers' blueprints. Parallel developments occurred simultaneously in biology, where researchers emphasized ecology, the context of organisms' lives, and organismic interactions with their environment. They demonstrated the results in diagrams, such as the familiar ones showing the water cycle from rain to evaporation to clouds to rain, or plants breathing carbon dioxide in and oxygen out that animals then take in. New Archaeologists sought data from paleobotanists and soil scientists to identify the climate changes that they wanted to correlate with cultural changes, an approach later tagged ecological determinism. Their commitment to natural history rather than culture histories persuaded them that human societies *must* follow evolutionary pathways. But by premising *unilinear* cultural evolution, New Archaeologists failed to understand evolutionary biology. That field of science has always, from Darwin through the twentieth-century luminaries Ernst Mayr and Stephen Jay Gould, demonstrated *diversity* of evolutionary paths, a fundamental aspect of natural selection. Segments of a parent population experience diverse ecologies, through environmental changes or by migration, and can diverge in a variety of directions, including regressive ones.

Pushed to the side during the New Archaeologists' campaign to dominate American archaeology was the multilineal model presented by Julian Steward, an anthropologist who had done archaeological work in Utah in the 1930s but was better known for his ethnography of Paiute and Shoshone in the desert West. In the 1950s, Steward wrote his *Theory of Culture Change*, proposing that cultures have core components tied to their adaptation to their environment, and a less essential layer of traits added by historical relationships of intersocietal contacts and occasional creative society members. The Paiute and Shoshone, for example, managed to find enough food to survive even in the sagebrush desert of Nevada. Beyond subsistence, they created highly artistic baskets; their art, songs, and stories, like those of the Inuit Boas lived with, were widely shared and greatly enhanced their lives. Steward's theory, given with illustrative cases, fit evolutionary biology better than did New Archaeology, but his multilineal model did not trump the less scientific unilineal scheme. Steward himself was engaged in an ambitious socio-cultural anthropological study of Puerto Rico rather than archaeology, so he did not push his model in archaeology conferences or reach many archaeology graduate students.

The postwar generation's commitment to discovering universal regularities in human societies became passé in the 1980s. The next generation proclaimed New Archaeology's blinkered methodology to be unsophisticated, naive about the limitations of the archaeological record and about the unpredictability of human behavior. Because the generation of the 1960s and 1970s called its archaeology "processual" (looking at ongoing processes rather than at events), the next generation were—of course—"postprocessualists." Their title aligned them with the concurrent mode in humanities studies, "postmodernism." Postmodernists highlighted biases arising from scholars' unexamined early personal socialization—for example, the assumption that only men did important work. There could be multiple versions of history depending on the standpoint and biases of the writer. Objectivity, said a historian analyzing famous earlier historians' efforts, was only "a noble dream." Postmodernism encouraged writers to describe their private feelings, use the first-person "I" instead of the objective third-person "outside narrator." An early example is archaeologist Janet Spector's *What This Awl Means*, which describes her awakening realization that the site she was excavating was once a real community of Dakota people. She related how she found their descendants, spoke with them, and learned that a simple bone sewing awl had been as treasured a possession for a Dakota woman as Spector's grandmother's thimble had been in her family.

After more than a decade of this sliding away from the old ideal of scientific rigor, some anthropologists and archaeologists tried to organize a movement within the discipline to reinstate scientific anthropology. If there can be multiple versions of history, there can be more than one correct model of objective science. Those folks who are committed to what they believe to be strictly empirical data now must live with colleagues pursuing gender ideology and roles in the past (see Chapter 5), colleagues trying to identify cognitive behavior from archaeological material, colleagues working with indigenous nations' interests and understandings of their pasts (Chapter 4), and the remaining no-longer-new New Archaeologists.

Recent years have seen the development of evolutionary archaeology, purporting to be Darwinian: natural selection for fitness eliminated some artifacts, and reproduced others. Humans apparently were artifact evolution's unwitting instruments. Following the strategy of Lewis Binford and his New Archaeologists a generation before, evolutionary archaeology leaders published prolifically, organized sessions at national meetings, and denounced everyone else as poor scientists. The second time around, a new "new archaeology" was less successful; too many other archaeologists were trying the other tack, to ascertain behavior of individuals in the record of the prehistoric past.

Thus, a variety of models are operating in the contemporary world of archaeological theory. Added to this stew is the development of cultural resources management (CRM), a contract archaeology business adhering to standardized procedures and legal regulations for the protection of heritage sites. This brand of archaeology is generally atheoretical, a throwback to the cultural historians who observed, collected, and analyzed data to tell us who lived where, and how, in the past. In this case, the conclusions are fashioned into CRM reports submitted to government agencies and corporations.

Cultural evolution, Elman Service's schema, did not disappear. Band-tribe-chiefdom-state unilinear evolution is embedded in Western culture, explicitly formulated in the late seventeenth century, described in the "conjectural histories" of the eighteenth-century Enlightenment, and taken for granted in the nineteenth century. No matter how glaring the racism inherent in slotting societies into Service's stages of cultural evolution and implying Progress toward civilization, many American archaeologists didn't see it as racism. Their respected professors had taught them the schema. Lacking protracted field experience with non-Western communities (such as Franz Boas had among the Inuit, which taught him how false it was to put Inuit into a lower stage of cultural evolution), the majority of archaeologists continued the traditional paradigm. A few years into the twenty-first century, the ground could now be shifting. An archaeologist who had graduated from the University of Michigan, dutifully learning band-tribe-chiefdom-state (although Leslie White had long ago died and Elman Service retired), tried to deal with Cahokia as a chiefdom. He quickly realized the model was inadequate and argued against it with a book entitled *Chiefdoms and Other Archaeological Delusions* (Pauketat 2007).

The steadily increasing number of archaeologists working with, often working for, First Nations simply cannot think of these colleagues' histories in the form of Western pseudo-evolutionary conjectural history. The only reasonable way to handle multiple histories is the way biologists understand the multiplicity of species, each with its particular evolutionary history. For humans, evolutionary histories of societies also include intersocietal contacts. Humans tend to travel; a desire to see more of the world seems to be part of *Homo sapiens'* nature. Between changes in the natural and political environments that must be adjusted to, and the curiosity more characteristic of humans than even of cats, human societies are never, and can never be, static. A single model of human social organization is too simplistic to encompass this diversity.

■ ■ ■ How Did Complex Societies Emerge from the Hunter-Gatherers that Came before Them?

These various models of cultural development clash when we attempt to theorize how complexity developed in human societies. About 12,000 years ago, all humans lived in small bands and subsisted through hunting, fishing, and gathering plant foods. Today, hunting-gathering communities have almost disappeared, and highly organized and technical cities hold tens of millions of people. How do archaeologists explain that?

Lewis Henry Morgan, the nineteenth-century pioneer American anthropologist discussed earlier in this chapter, worked up a scheme for cultural evolution using a magic formula of three times three (three is the magic number for most Indo-European speakers) stages in a unilineal cultural evolution, a predecessor to Service's grouping. This scheme is shown in Table 10.1.

Notice that Morgan, although not an archaeologist, lists imperishable *material culture* signs of evolutionary progress. His sequence didn't hold up once stratigraphic excavations, and then radiocarbon dating, revealed that pottery appeared earliest in the terminal Pleistocene in northeastern Asia, several millennia later in western and southern Eurasia and Latin America; that agriculture preceded pottery in the Near East; that bows and arrows appeared at the end of the Pleistocene in Europe but not until mid-first millennium A.D. in the Americas; and that some highly complex societies never developed writing systems. Morgan's singling out of irrigation agriculture survived into the mid-twentieth century, as Julian Steward built his theory of the development of states upon it.

Steward followed a Marx-influenced historian (married to an anthropologist) to postulate that in arid lands, farmers would have cooperated to dig ditches to bring river water to their fields. Growing populations, fed by the fields, expanded systems of ditches until it became necessary to appoint someone to manage them, organizing farmers to maintain the ditches and allocating water equably along the system. In time, the scenario said, managers would become

Table 10.1 Morgan's Scheme for Cultural Evolution (read from bottom, oldest, to top [present time])

Civilization	From the invention of a phonetic alphabet, with the use of writing, to the present time.
Upper Status of Barbarism	From invention of smelting iron and iron tools
Middle Status of Barbarism	From domestication of animals (Old World) to cultivation of maize and other plants by irrigation, and adobe-brick and stone construction (Americas)
Lower Status of Barbarism	From invention of the art of pottery
Upper Status of Savagery	From invention of the bow and arrow
Middle Status of Savagery	From acquisition of a fish subsistence and knowledge of the use of fire
Lower Status of Savagery	From the infancy of the human race

(Source: Morgan 1985 [1877]:12–13)

despotic, knowing they held life and death power over the farmers through giving, or withholding, precious water. Thus evolved states. The scenario seemed to work in Mesopotamia, Egypt, China, India, and Peru, at least insofar as irrigation agriculture was the lifeblood of the states. It spurred major archaeological projects to actually find out which came first, irrigation agriculture or states. Empirical data punched holes in the theory: generally, early states in the great river valleys don't seem to have taken over village irrigation systems, and many of the water-control works were designed to retain annual floodwaters, rather than to lead river water to fields.

So why and how did some human societies grow to be complexly organized and dictatorial? And the reverse of that question: Why did other societies *not* evolve into states with cities? New Archaeologists pointed to environment plus demography (population size and density). States could evolve only where there was enough agricultural potential to feed large, dense populations. Population would increase to the point that authorities had to be appointed, occupational specialties encouraged, and religious and military orders developed. Societies that did not evolve into states had been held back by inadequate natural resources. How then to explain the empires created by steppe nomads in Eurasia—Scythians, Huns, Mongols, Ottoman Turks—groups that never relied on agriculture? How to explain rich farming regions that didn't develop cities and states, like the Iroquois in New York, or the pre-contact Midwest after the collapse of Cahokia? Was Chaco culture, located in a New Mexico arid valley with a meager river that couldn't support a large population, a state for a couple of centuries? Ecological determinism fell short as a universal explanation.

Postprocessualists rejected the search for universal laws, regularities, or explanations as foolish or misguided. When you came down to it, what is a state? Can you have cities without states? States without cities? Is Service's band-tribe-chiefdom-state model an arbitrary Western academic classification scheme? Women archaeologists were bothered by unilineal evolutionists' implicit focus on men's important business, neglecting households and gender roles as unimportant. Half of the human species should not be overlooked. Carole Crumley broke through the unilineal scheme by arguing that hierarchy isn't an inevitable aspect of larger societies. She used the term *heterarchy* to describe societies with a number of leaders or perhaps self-governing groups. Hierarchical societies have pyramid-shaped, stratified social ranks, with a ruler on the top (originally, the word *hierarch* referred to a chief priest). Heterarchies are the other (*hetero-*, "other" or "different") way societies can be organized, with different segments sharing power with the others. Farmers sharing an irrigation system can negotiate equitable distributions of water without depending on the state; they can negotiate individually or in groups, and if disputes are not easily resolved, the group can call in a mediator. Crumley was familiar with informal self-governing groups such as jump-rope play groups among young girls, whereas men were socialized in boyhood to be competitive and build hierarchies.

Heterarchy helped other archaeologists see alternative interpretations of data. Why did the Iroquois not evolve into a state? They evolved a heterarchical system of autonomous extended-family households represented in democratic village councils, which in turn sent representatives to the alliance of towns called the Haudenosaunee (League of the Iroquois). At each level, governance was through consensus decisions; it could take days of discussion before disagreements were resolved. Benjamin Franklin observed the Iroquois form of governance in action during treaty negotiations between Pennsylvania and Iroquois nations on its borders. Years later, he insisted at the United States Constitutional Convention that we did not need a rigid hierarchical government, that a federal structure with a democratic election of representatives could reach a workable consensus. Economists also are familiar with heterarchical systems in the market, where producers, shippers, and consumers often negotiate reasonably mutually beneficial terms without external interference. The music scene is another heterarchical system, band members working out among themselves how they will play, and then as a band, negotiating for concert venues.

Several archaeological publications early in the twenty-first century directly challenged the conventional notion of a hierarchical band-tribe-chiefdom-state, already rejected by Crumley in the previous decade. Looked at anew, Mesopotamia did not gradually get larger and larger settlements until city size was reached, nor did tribal chiefs gradually get richer and more authoritative. Instead, agricultural communities went on reproducing themselves for four or five thousand years; then, within a few centuries, some grew rapidly in space and population, built larger temples, palaces, and administrative compounds, and invented writing (around 3500 B.C.E.). Their texts tell us that the villages remaining around the city became integrated into the urban political and economic structure. At first glance, this system seems hierarchical, especially if one naively believes the boasts of kings who left many of the written records preserved to us. Mesopotamian city-states retained considerable heterarchy in that temples, kings, noble estate-owners, and merchants each had their organizations and had to accommodate to the others. Archaeologists thinking hard about what happened in Mesopotamia around 3000 B.C.E. realized that the city-states were really not so much "more complex" than they were efforts to *simplify* relationships through well-articulated principles of law and convention. Commerce, tenant-landowner contracts, military duties, police and justice were regularized for better overall management. Regulation fostered expansion by reducing conflict and providing security. Once one or a few communities consolidated into city-states, other communities with which they traded or contested territory rapidly organized in response.

Dropping the academic construct band-tribe-chiefdom-state has been particularly illuminating for archaeologists working in central Africa. Sub-Saharan Africa was overrun in the later nineteenth century by European imperial nations—Britain, France, Germany, Belgium—using rifles and artillery against states unable to arm themselves with comparable weapons. Black Africans were called primitive, said to be barely evolved from the apes in their jungles. It was, of course, quite legitimate, according to nineteenth-century European modes of thought, to conquer and exploit primitive peoples. In consequence of this cant of conquest, archaeologists did not expect to find cities. The structure of the cultures they did find seemed anomalous until the concept of heterarchy was applied. Along the Niger River in Mali, for example, there are a great number of villages in clusters, going back three thousand years. Many specialized in selected occupations, as potters, herders, grain farmers, or market towns. A cluster coordinated its relatively large population without building tightly packed residences within city walls, yet it functioned as a city. Niger River Valley people, like North American First Nations, avoided the heavy dependence on a few crops that brought famine to Europe and China. They remained skilled in harvesting native foods that didn't need cultivation; they fished, and hunted, and also cultivated new crops. Their river, the Niger, does not flow through arid country like Egypt and Mesopotamia, although it does, like the Nile, Tigris, and Euphrates, flood annually, depositing rich silt. It meanders and switches channels readily, requiring Malian farmers to continually adjust their landholdings, perhaps a factor inclining them to cluster villages rather than consolidate into a European-style city.

The early-twenty-first century critique of the conventional band-tribe-chiefdom-state model is a nice example of science historian Thomas Kuhn's concept of paradigms established in normal science being overturned by the accumulation of data initially perceived as anomalies (see Chapter 2). Two breakthroughs changed thinking about the development of complex societies: Crumley's essays arguing for both hierarchical and heterarchical forms of societal structure, and the realization that complex states' governance systems were designed to simplify interactions between people through standardization. This internal development of greater sophistication within the discipline of archaeology in recent decades is surely influenced by external shifts, as the nations of the world have moved from a hierarchical organization of empires with colonies, to a heterarchical United Nations, European Union, and free-market paradigm. It takes time for a newer paradigm to diffuse, time for an older generation to die out, and younger people socialized in the new paradigm to rise into positions of power. The lag is exemplified in the following case.

Delgamuukw

The official line on American First Nations during the nineteenth century was published by the Director of the Smithsonian's Bureau of American Ethnology, Civil War veteran Major John Wesley Powell:

> When civilized man first came to America the continent was partially occupied by savage tribes, who obtained subsistence by hunting, by fishing, by gathering vegetal products, and by rude garden culture in cultivating small patches of ground. . . . The attempts to educate the Indians and teach them the ways of civilization have . . . disappointed their enthusiastic promoters. . . . The great boon to the savage tribes . . . has been the presence of civilization, which, under the laws of acculturation, has irresistibly improved their culture by substituting new and civilized for old and savage. (Powell 1881:xxvii–xxx)

Now, when do you suppose the next statement was made?

> It is common, when one thinks of Indian land claims, to think of Indians living off the land in pristine wilderness. . . . . there is no doubt, to quote Hobbs [sic], that aboriginal life in the territory was, at best, "nasty, brutish and short" (p. 236) . . . [American Indians'] unsettled habitation . . . cannot be account-

> ed a true and legal possession, and the people of Europe . . . finding land of which the Savage stood in no particular need, and of which they made no actual and constant use, were lawfully entitled to take possession (p. 239). . . . Some tribes are so low in the scale of social organization that . . . they likely acted as they did because of survival instincts. (McEachern, *Reasons for Judgment*, quoted in Culhane 1998:236, 239, 247, 248).

Answer: In British Columbia in 1991 when Chief Justice Allan McEachern of the British Columbia Supreme Court ruled, in Delgamuukw v. Regina,[2] on a claim for territorial rights by the Gitxsan (Figure 10.1) and Wet'suwet'en First Nations of that province. "It is my conclusion," he said, "that Gitksan and Wet'suwet'en laws and customs are not sufficiently certain to permit a finding that they or their ancestors governed the territory" (*ibid.*:249).

The plaintiff Indian nations appealed McEachern's decision to the British Columbia Court of Appeal, which in 1993 ruled that McEachern had erred in finding extinguishment of aboriginal title. Then in December 1997, the

FIGURE 10.1 'Ksan village, Gitxsan Tsimshian Nation, British Columbia. This is the territory claimed in the Delgamuukw case. *Photo by Alice B. Kehoe.*

---

[2] Delgamuukw is the title of the hereditary leader of a Gitxsan lineage; "Regina" means "the Crown," the Queen of the British Commonwealth, formally the defendant in cases against the government in Canada.

Supreme Court of Canada ruled that McEachern erred in refusing to give weight to the plaintiffs' formal oral histories, Gitxsan *adaawk* and Wet'suwet'en sung *kungax* (see Chapter 4), and ordered a new trial. At that point, the two small, weary nations agreed to negotiate a settlement with the province.

Justice McEachern's judgment proves the tenacity of nineteenth-century racist doctrines in Anglo America. The United States had similarly ruled, in 1955, in *Tee-Hit-Ton v. United States*, that First Nations had no legal title to their territories and therefore, the United States does not owe them any compensation for taking their land! U.S. Supreme Court Justice Stanley Reed stated, in that case:

> Every American schoolboy knows that the savage tribes of this continent were deprived of their ancestral range by force and that, even when the Indians ceded millions of acres by treaty in return for blankets, food, and trinkets, it was not a sale but the conquerors' will that deprived them of their land. (quoted in Wilkins and Lomawaima 2001:24)

Both the Tee-Hit-Ton, a Tlingit community in Alaska, and the Gitxsan and Wet'suwet'en were particularly interested in harvesting timber from the aboriginal territories around their present villages, rather than watch commercial non-Indian logging companies cut down neighboring forests without giving them any compensation. In each case, the respective first Court decisions deprived small communities of one of their few sources of income, by arguing *not from legal precedent* but from alleged anthropological knowledge. Western cultural tradition supporting imperial conquest and appropriation was legitimated by the myth of savages in a wilderness, bands-tribes-chiefdoms in lower stages of cultural evolution. We can't blame contemporary anthropologists, because their testimonies were rejected in these cases; Justice McEachern specifically barred anthropologists from contributing on grounds that, having lived and worked with the plaintiffs, they would be biased in their favor. Archaeological work in the disputed area of British Columbia was dismissed as inconclusive because it could not substantiate the existence of specifically Gitxsan and Wet'suwet'en wooden lineage houses two thousand years ago. The archaeologist's data from which he inferred stratified societies owning resources and territories in the region as early as 500 B.C.E. were dismissed as too indefinite.

Maybe the era of benighted would-be civilizers has passed. Canada's Supreme Court did, after all, bring the twentieth century to a close by invalidating McEachern's judgment. The United States Indian Self-Determination Act, in 1975, reversed long-standing absolutist domination of Indian tribes. NAGPRA admits First Nations' interests in prehistoric remains. Canada's 1998

Statement of Reconciliation with its First Nations announced a policy of rectifying injustices. We anthropologists have sworn off using the word *primitive*.

Controversies will continue, but there seems to be a heartening reaffirmation of the importance of empirical data. There was a real past out there, only most of it disappeared, and what's left will be interpreted by a diversity of people, from a variety of backgrounds and inclined to emphasize some data more than others. If we can't know it all, we can keep our minds curious and our hands on the trowel, building chains of signification to inferences to the best explanations.

SOURCES

Adams, Henry, 1918. *The Education of Henry Adams*. New York: Random House.

Crumley, Carole L., 1987. A Dialectical Critique of Hierarchy. In *Power Relations and State Formation*, edited by Thomas C. Patterson and Christine W. Gailey, pp. 155–169. Washington, DC: Archeology Section/American Anthropological Association.

—. 1995. Heterarchy and the Analysis of Complex Societies. In *Heterarchy and the Analysis of Complex Societies*, edited by Robert M. Ehrenreich, Carole L. Crumley, and Janet E. Levy, pp. 1–5. Archeological Papers, No. 6. Washington, DC: American Anthropological Association.

Culhane, Dara, 1998. *The Pleasure of the Crown: Anthropology, Law, and First Nations*. Burnaby, B.C.: Talonbooks.

Gould, Stephen Jay, 2002. *The Structure of Evolutionary Theory*. Cambridge, MA: Belknap.

Kehoe, Alice Beck, 1998. *The Land of Prehistory: A Critical History of American Archaeology*. New York: Routledge.

Mayr, Ernst, 1982. *The Growth of Biological Thought: Diversity, Evolution, and Inheritance*. Cambridge, MA: Belknap.

McIntosh, Roderick J., 2005. *Ancient Middle Niger: Urbanism and the Self-Organizing Landscape*. Cambridge: Cambridge University Press.

Morgan, Lewis Henry, 1985 [1877]. *Ancient Society*. Tucson: University of Arizona Press. Facsimile.

Novick, Peter, 1988. *That Noble Dream: The "Objectivity Question" and the American Historical Profession*. Cambridge: Cambridge University Press.

Pauketat, Timothy R., 2007. *Chiefdoms and Other Archaeological Delusions*. Lanham, MD: AltaMira Press (Rowan and Littlefield).

Phillips, David A., Jr., and Lynne Sebastian, editors, 2001. *Exploring the Course of Southwest Archaeology: The Durango Conference, September 1995*. Albuquerque: New Mexico Archaeological Council.

Powell, John Wesley, 1881. Report of the Director. First *Annual Report*, Bureau of Ethnology, 1879–80, pp. xi–xxxiii. Washington, DC: Government Printing Office.

Smith, Adam T., 2003. *The Political Landscape: Constellations of Authority in Early Complex Polities*. Berkeley: University of California Press.

Spector, Janet, 1993. *What This Awl Means: Feminist Archaeology at a Wahpeton Dakota Village*. St. Paul: Minnesota Historical Society Press.

Steward, Julian H., 1955. *Theory of Culture Change*. Urbana: University of Illinois Press.

Trigger, Bruce A., 1998. *Sociocultural Evolution*. Oxford: Blackwell.

Wilkins, David E., and K. Tsianina Lomawaima, 2001. *Uneven Ground: American Indian Sovereignty and Federal Law*. Norman: University of Oklahoma Press.

Yoffee, Norman, 2005. *Myths of the Archaic State: Evolution of the Earliest Cities, States, and Civilizations*. Cambridge: Cambridge University Press.

Notes

To help students to focus on concepts instead of memorizing lists of names, I have omitted many archaeologists' names from this text. For instructors who may be unfamiliar with some of the research areas discussed, this key is provided. See also authors cited in the "Sources" at the end of each chapter.

Introduction

p. 14 "surprising research on butchered mammoths": David F. Overstreet. I refrain from naming the man in the tight black t-shirt.

Chapter 2

p. 40 "Outstanding historian": Samuel Eliot Morison.
p. 46 "cultural anthropologist . . . who read the paper": Luke Eric Lassiter.
p. 54 "a wordsmith": Dody H. Giletti.
p. 54 "archaeologist of the younger generation": Dale Walde.

Chapter 3

p. 59 95-year-old archaeologist: W. Curry Holden; his "graduate student," Jane Holden Kelley (his daughter, then a graduate student in anthropology at Harvard).
p. 62 "newly discovered cave"—allegedly, "Burrows Cave," promoted by Russ Burrows and Frank Joseph (Collin).
p. 65 Moundbuilders study: Cyrus Thomas headed the research project.
p. 68 "Reinhard": Reinhard Lehne was so good at interpreting archaeological data, using experience in lieu of formal education, that he was hired to work full-time in the archaeological laboratory during the winters.
p. 74 "one British archaeologist argued": Colin Renfrew, now Lord Renfrew.

Chapter 4

p. 91 "Women archaeologists who worked also as ethnographers": Besides de Laguna, Florence Hawley Ellis, and myself, twentieth-century examples include Ruth Gruhn, Jane Holden Kelley, Isabel Kelly, Susan Kent, Dorothy Keur, Clara Lee Fraps Tanner, Patty Jo Watson. Among men, there is Thomas Dillehay, who has been working with Mapuche as well as excavating Monte Verde in Chile. This list excludes nineteenth- and turn-of-the-twentieth-century anthropologists such as Frank Hamilton Cushing and Alfred Kroeber who as a matter of course worked broadly in anthropology.

p. 92 "Blackfeet Tribal Historic Preservation Office hired an archaeologist": María Nieves Zedeño. The Tribe also had a busy staff archaeologist, Betty Matthews, who is Blackfoot.

p. 94 "non-Indian anthropologist intervened on behalf of Omaha friends": Robin Ridington.

p. 94 "archaeologists who can't believe oral histories": Ronald J. Mason argued this position in his 2006 book, *Inconstant Companions* (University of Alabama Press).

p. 94 "Osage": See Chapter 9, section on interpreting Cahokia with Osage texts collected by Smithsonian ethnographer Francis La Flesche.

p. 95 "Thomas Jefferson deliberately lied": Historian Peter S. Onuf describes Jefferson as "uncompromising, self-righteous, and dangerously doctrinaire." See Onuf 1997:141. See also Wallace 1997 and 1999.

p. 97 "Then came archaeology." After a grass fire burned over the Little Bighorn battlefield in 1983, the National Park Service superintendent for the site invited an archaeologist, Richard Fox, to take advantage of the lack of vegetation cover to search for artifacts and other evidence. National Park Service staff archaeologist Douglas Scott then collaborated with Fox on a two-year field project. The book, listed in "Sources," with Scott as first author, is the final report on the project. Fox's book, also in "Sources," includes more discussion of his methodological principles, and more on using Indian accounts. Gregory F. Michino (Michino 1997) uses Fox's work and a time-and-motion study of movement on the battlefield by John S. Gray to tell the final battle in ten-minute intervals. A convenient compendium of Indian accounts is by Richard G. Hardorff (Hardorff 1991).

Chapter 5

p. 102 "$40 on projects that didn't hire women": The Smithsonian's River Basin Surveys, largest employer of field crews in the 1950s.

p. 102 footnote 1: See, e.g., Barber 1952.

p. 103 "Indian Claims Commission": The Indian Claims Commission Act was drafted by Felix S. Cohen, a lawyer who wrote most of the Indian New Deal legislation of Franklin Roosevelt's administration during the late 1930s.

p. 105 "bone tools for evidence for basketry and weaving": See my paper, Kehoe 1991a.

p. 106 cemeteries in Wisconsin: See Pleger 2000.

p. 107 "Veblen": See my paper, Kehoe 1999.

p. 108 "brothels": See Seifert 1991.

p. 112 "Marxist-inspired" Ludlow Project: Randall McGuire, director, has published *A Marxist Archaeology* (McGuire 1992). The paper referred to is by Margaret Wood (Wood 2002).

p. 114 "A Philadelphia civic leader": Michael Coard, quoted in Associated Press story, *Milwaukee Journal Sentinel*, June 9, 2007.

p. 114 "cemetery [of] eighteenth-century African-Americans": Biological anthropologist Michael Blakey remarked, "his forced labors were backbreaking in the most literal sense," quoted in Wall and Cantwell 2004:52.

p. 116 "California archaeologists . . . dismissive": Alfred Kroeber and Robert Heizer led this conservative view.

p. 117 "successions of empires": See Herzfeld 1992.

Chapter 6

p. 120 "Don't confuse my mind with facts." Or as comedian Richard Pryor put it, "Who are you going to believe—me, or your lying eyes?" Quoted by Henry Louis Gates, Jr., *New Yorker* Oct. 23, 1995, p. 64.

p. 123 Çatalhöyük excavator: In the 1960s, James Mellaart; beginning in the 1990s, Ian Hodder.

p. 125 "South African studying rock art": The instigator of the shaman-in-trance interpretation was David Lewis-Williams. He initially published with Thomas Dowson. David Whitley is an American disciple, and Jean Clottes is the French archaeologist applying Lewis-Williams's ideas to Paleolithic cave art. Archaeologist opponents of Lewis-Williams's interpretations include, beside myself, Paul G. Bahn and Anne Solomon.

p. 126 "French anthropologist witnessing . . .": Roberte Hayamon.

p. 128 "astute British anthropologist": E. E. Evans-Pritchard.

p. 128 "neurophysiologist": Patricia Helvenston. She has co-authored papers with Paul Bahn. They have critiqued the Khoisan ethnography; see Helvenston and Bahn 2006.

p. 137 "conflict between science and religion": There are a number of books on the subject, a few, as those by Richard Dawkins, with a negative opinion on religion, but most asserting that there need be no conflict. In the 1980s, Pope John Paul II accepted a German Catholic theologian's view that Christians *should* accept evolutionary biology as an expression of God's plan, and issued encyclicals advising Roman Catholics that their Church follows this scientific theory. See also Petto and Godfrey 2007. My chapter in this volume discusses the history of the alleged conflict.

Chapter 7

p. 144 "Manifest Destiny": See my *Land of Prehistory* (Kehoe 1998).
p. 145 "bold example by two Europeans": Kristiansen and Larsson, *Rise of Bronze Age Societies* (2005).
p. 148 "Iñupiaq": For more information and sources on North American Indians including indigenous Mexico (Olmec, below), see my book *North American Indians* (Kehoe 2006).
p. 160 "forensic chemist": Svetlana Balabanova (Balabanova, Parsche, and Pirsig 1992) and subsequent reports on analyses of nicotine, cocaine, and hashish in skeletons (e.g., Jett 2002)
p. 160 "maize in South India, etc.": See Sorenson and Johannessen 2004 and 2006.
p. 166 "Richard Nielsen replied, 'Minnesota'": Garrison Keillor, proud Minnesotan, the Bard of Lake Wobegon, described the 1362 arrival of Norse in Kensington as Chapter 1 of "The Story of Minnesota," broadcast August 12, 2006, on "The Prairie Home Companion."
p. 165 "Eight Götlanders": Götland is a district and an island in Sweden.
p. 166 "flakes from stone tools": Scandinavian farmers knapped stone tool blades until the seventeenth century A.D., so the flakes could have been knapped by Norsemen.
p. 167 "Hudson's Bay Company": By the 1660s, Britain's prosperity gave the aristocrat class considerable wealth they wanted to invest. The Hudson's Bay Company, in the north, and Carolina Colony, in the south, were well capitalized by these investors; both returned profits through trading animal skins—furs in the north, deer hides in the south.
p. 167 "paleo-geographer": Prof. Reid Bryson, University of Wisconsin.
p. 168 "weight of probability": Prof. Guy Gibbon, archaeologist at the University of Minnesota, suggested this metaphor.

Chapter 8

p. 175 "astronomer proposed . . . moon eclipses": Gerald Hawkins.
p. 178 "our astronomer collaborator [Moose Mountain]": John A. Eddy.
p. 178 "avocational archaeoastronomer . . . confirmed": Allan G. Fries.
p. 179 "rendezvous": I have argued since 1981 (Kehoe 1981) that these annual multi-band rendezvous were intermittent polities, fulfilling the functions of city-states outlined by Aristotle; see Chapter 10 on the multiplicity of societal evolutions.
p. 190 "Africa to Brazil": seriously proposed by the reputable, if unconventional, archaeologist Donald Lathrap in the 1970s: See Lathrap 1977.
pp. 190, 192 "along the North Atlantic islands": Dennis Stanford and Bruce Bradley, reviving an idea proposed by Emerson Greenman. Stanford and Bradley argue the European forebears were Solutrean (see later in this chapter).
p. 192 "Gryba": Personal communication, several discussions.
pp. 192–93 "Meadowcroft's excavator": James Adovasio.
p. 193 "Chesrow": Sites excavated by David Overstreet. See Overstreet 1988.

Chapter 9

p. 208 "Hobbes": He wrote *Leviathan* first in English, later translated it into Latin, and made "where every man is enemy to every man" into "bellum omnia contra omnes," "war of all against all."

p. 210 "two well-respected archaeologists": Jonathan Haas and David Wilcox.

p. 210 "Bryan": See Numbers 1992.

Chapter 10

p. 220 "vital force": Herbert Spencer, prolific Victorian writer. The best reference is Peel 1971.

p. 220 "Superorganic": See Kroeber, "The Superorganic," reprinted in Kroeber 1952.

p. 221 "agency": The standpoint that we must seek evidence of individuals and their actions.

p. 221 "realpolitik": German for "real politics," meaning actuality rather than rhetoric and idealism.

p. 223, footnote 1 "radical woman": Victoria Claflin Woodhull (1838–1927), first woman to run, in 1872, for President of the United States (although women could not vote), first woman stockbroker, and a spiritualist—she had an astounding life. See Goldsmith 1998 or Gabriel 1998.

p. 225 "New Archaeology": A gossipy history is *Archaeology as a Process* (O'Brien, Lyman, and Schiffer 2005).

p. 227 "evolutionary archaeology": Michael O'Brien and Lee Lyman developed this approach from ideas taught by Robert Dunnell. Douglas Bamforth critiqued it in *American Antiquity* (Bamforth 2002), and I had done so in *Review of Archaeology* (Kehoe 2000c).

p. 227 "*Chiefdoms and Other Archaeological Delusions*": Timothy Pauketat describes how his empirical work at Cahokia and its environs led him to see that the "Michigan School" band-tribe-chiefdom-state paradigm could not adequately account for his data.

p. 228 "Marx-influenced historian": Karl Wittfogel, married to Boas student Esther Goldfrank. Wittfogel studied the economic history of Asia, where irrigation was highly developed.

p. 231 "Looked at anew, Mesopotamia": Adam Smith's (2003) and Norman Yoffee's (2005) books both focus on the authors' research in Mesopotamia and among its neighbors. Roderick McIntosh and his wife Susan McIntosh, also an archaeologist, critique standard cultural-evolutionary models from the standpoint of their West African data. See McIntosh 2005.

p. 234 "biased in their favor" and "archaeological work": See Culhane 1998:156–159.

Bibliography

Adams, Henry. 1931 [1918]. *The Education of Henry Adams*. New York: Modern Library.

Anderson, M. Kat. 2005. *Tending the Wild: Native American Knowledge and the Management of California's Natural Resources*. Berkeley: University of California Press.

Anthony, David W. 2007. *The Horse, the Wheel, and Language: How Bronze-Age Riders from the Eurasian Steppes Shaped the Modern World.* Princeton: Princeton University Press

Arkush, Elizabeth N., and Mark W. Allen, editors. 2006. *The Archaeology of Warfare: Prehistories of Raiding and Conquest*. Gainesville: University of Florida Press.

Arnold, Bettina. 1990. The Past as Propaganda: Totalitarian Archaeology in Nazi Germany. *Antiquity* 64:464–478.

—. 1991. The Deposed Princess of Vix: The Need for an Engendered European Prehistory. In *The Archaeology of Gender: Proceedings of the 22nd Annual Chacmool Conference*, edited by Dale Walde and Noreen D. Willows, pp. 366–374. Calgary: Archaeological Association of the University of Calgary.

—. 1998. The Power of the Past: Nationalism and Archaeology in 20th Century Germany. *Archaeologia Polonia* 35–36:237–253.

Aveni, Anthony F. 2001. *Skywatchers*. Austin: University of Texas Press.

Balabanova, Svetlana, F. Parsche, and W. Pirsig. 1992. First Report of Drugs in Egyptian Mummies. *Naturwissenschaften* 79:358.

Bamforth, Douglas. 2002. Evidence and Metaphor in Evolutionary Archaeology. *American Antiquity* 67 (3):435–452.

Barber, Bernard. 1952. *Science and the Social Order*. Glencoe, IL: Free Press.

Barnes, Barry, David Bloor, and John Henry. 1996. *Scientific Knowledge: A Sociological Analysis*. Chicago: University of Chicago Press.

Barton, C. Michael, Geoffrey A. Clark, David R. Yesner, and Georges A. Pearson, editors. 2004. *The Settlement of the American Continents*. Tucson: University of Arizona Press.

Bellwood, Peter. 2005. *First Farmers: The Origins of Agricultural Societies*. Malden, MA: Blackwell.

Bernardini, Wesley. 2005. *Hopi Oral Tradition and the Archaeology of Identity*. Tucson: University of Arizona Press.

Bernal, Martin. 1991. *Black Athena: The Afroasiatic Roots of Classical Civilization*, Vol. II. New Brunswick, NJ: Rutgers University Press.

Binford, Lewis R. 1968. Archeological Perspectives. In *New Perspectives in Archeology*, edited by Sally R. Binford and Lewis R. Binford, pp. 5–32. Chicago: Aldine.

Boehm, David A., Stephen Topping, and Cyd Smith, editors. 1983. *Guinness Book of World Records*. New York: Sterling.

Brodie, Neil, Morag M. Kersel, Christina Luke, and Kathryn Walker Tubb, editors. 2006. *Archaeology, Cultural Heritage, and the Antiquities Trade*. Gainesville: University Press of Florida.

Brownlee, Kevin, and E. Leigh Syms. 1999. *Kayasochi Kikawenow: Our Mother from Long Ago (An Early Cree Woman and Her Personal Belongings from Nagami Bay, Southern Indian Lake)*. Winnipeg: Manitoba Museum of Man and Nature.

Buikstra, Jane E., and Lane A. Beck, editors. 2006. *Bioarchaeology: The Contextual Analysis of Human Remains*. Amsterdam: Academic Press/Elsevier.

Carlson, John B. 1984. The Nature of Mesoamerican Astronomy: A Look at the Native Texts. In *Archaeoastronomy and the Roots of Science*, edited by E. C. Krupp, pp. 211–252. Boulder, CO: Westview Press.

Castillo Butters, Luis Jaime, and Ulla Sarela Holmquist Pachas. 2006. Modular Site Museums and Sustainable Community Development at San José de Moro, Peru. In *Archaeological Site Museums in Latin America*, edited by Helaine Silverman, pp. 130–155. Gainesville: University Press of Florida.

Charles, Douglas K., and Jane E. Buikstra, editors. 2006. *Recreating Hopewell*. Gainesville: University Press of Florida.

Clark, John E., and Michelle Knoll. 2005. The American Formative Revisited. In *Gulf Coast Archaeology: The Southeastern United States and Mexico*, edited by Nancy Marie White, pp. 281–303. Gainesville: University Press of Florida.

Crumley, Carole L. 1987. A Dialectical Critique of Hierarchy. *In Power Relations and State Formation*, edited by Thomas C. Patterson and Christine W. Gailey, pp. 155–169. Washington, DC: Archeology Section, American Anthropological Association.

—. 1995. Heterarchy and the Analysis of Complex Societies. In *Heterarchy and the Analysis of Complex Societies*, edited by Robert M. Ehrenreich, Carole L. Crumley, and Janet E. Levy, pp. 1–5. Archeological Papers, No. 6. Washington, DC: American Anthropological Association.

Culhane, Dara. 1998. *The Pleasure of the Crown: Anthropology, Law, and First Nations*. Burnaby, B.C.: Talonbooks.

De Laguna, Frederica. 1960. *The Story of a Tlingit Community: A Problem in the Relationship between Archeological, Ethnological, and Historical Methods*. Smithsonian Institution, Bureau of American Ethnology Bulletin 172. Washington, DC: Government Printing Office.

—. 2000. *Travels among the Dena: Exploring Alaska's Yukon Valley*. Seattle: University of Washington Press.

Dever, William G. 1990. *Recent Archaeological Discoveries and Biblical Research*, Seattle: University of Washington Press.

—. 1999. American Palestinian and Biblical Archaeology. In *Assembling the Past*, edited by Alice Beck Kehoe and Mary Beth Emmerichs, pp. 91–102. Albuquerque: University of New Mexico Press.

Dixon, Kelly J. 2005. *Boomtown Saloons*. Reno: University of Nevada Press.

Doran, Edwin, Jr. 1973. *Nao, Junk, and Vaka: Boats and Culture History*. College Station: Texas A&M University.

Eller, Cynthia. 2000. *The Myth of Matriarchal Prehistory: Why an Invented Past Won't Give Women a Future*. Boston: Beacon Press.

Evans-Pritchard, E. E. 1963. *Essays in Social Anthropology,* Glencoe, IL: Free Press.

Fell, Barry. 1976. *America B.C.* New York: Pocket Books.

Ferguson, T. J., and Chip Colwell-Chanthaphonh. 2006. *History Is in the Land: Multivocal Tribal Traditions in Arizona's San Pedro Valley*. Tucson: University of Arizona Press.

Fingerhut, Eugene R. 1994. *Explorers of Pre-Columbian America? The Diffusionist-Inventionist Controversy*. Claremont, CA: Regina.

Finkelstein, Israel, and Neal A. Silberman. 2006. *David and Solomon*. New York: Free Press.

Flenley, John, and Paul G. Bahn. 2003. *The Enigmas of Easter Island.* Oxford: Oxford University Press.

Ford, James A. 1969. *A Comparison of Formative Cultures in the Americas.* Smithsonian Contributions to Anthropology 11. Washington, DC: Smithsonian Institution Press.

Fowler, Melvin L., Jerome Rose, Barbara Vander Leest, and Steven R. Ahler. 1999. *The Mound 72 Area: Dedicated and Sacred Space in Early Cahokia*. Reports of Investigations, No. 54. Springfield IL: Illinois State Museum.

Fox, Richard G., editor. 1998. The Neanderthal Problem and the Evolution of Human Behavior. Special issue of *Current Anthropology* 39, Supplement.

Gabriel, Mary. 1998. *Notorious Victoria*. Chapel Hill, NC: Algonquin.

Gerstenblith, Patty. 2006. Recent Developments in the Legal Protection of Cultural Heritage. In *Archaeology, Cultural Heritage, and the Antiquities Trade*, edited by N. Brodie, M. M. Kersel, C. Luke, and K. W. Tubb, pp. 68–92. Gainesville: University Press of Florida.

Gholi Majd, Mohammad. 2003. *The Great American Plunder of Persia's Antiquities, 1925–1941*. Lanham, MD: University Press of America.

Goldsmith, Barbara. 1998. *Other Powers: The Age of Suffrage, Spiritualism, and the Scandalous Victoria Woodhull.* New York: Knopf.

Gould, Stephen Jay. 2002. *The Structure of Evolutionary Theory*. Cambridge, MA: Belknap.

Green, Roger C. 1998. Rapanui Origins Prior to European Contact: The View from Eastern Polynesia. In *Easter Island and East Polynesian Prehistory*, edited by P. Vargas Casanova, pp. 87–110. Santiago: Universidad de Chile, Instituto de Estudios Isla de Pascua.

Haas, Jonathan, and Winifred Creamer. 1997. Warfare among the Pueblos: Myth, History, and Ethnology. *Ethnohistory* 44 (2):235–61.

Hägerstrand, Torsten. 1967. *Innovation Diffusion as a Spatial Process*. Chicago: University of Chicago Press.

Hamdani, Abbas. 2006. Arabic Sources for the Pre-Columbian Voyages of Discovery. *Maghreb Review* 31 (3–4):203–221.

Hardorff, Richard G. 1991. *Lakota Recollections of the Custer Fight: New Sources of Indian-Military History*. Reprinted 1997 in paperback by University of Nebraska Press.

Harris, David R., editor. 1996. *The Origins and Spread of Agriculture and Pastoralism in Eurasia*. Washington, DC: Smithsonian Institution Press.

Harvey, Graham, editor. 2003. *Shamanism: A Reader*. London: Routledge.

Haynes, Gary. 2002. *The Early Settlement of North America*. Cambridge: Cambridge University Press.

Hays-Gilpin, Kelley. 2004. *Ambiguous Images: Gender and Rock Art*. Walnut Creek, CA: AltaMira.

Helvenston, Patricia, and Paul G. Bahn. 2006. Archaeology or Mythology? The "Three Stages of Trance" Model and South African Rock Art. *Cahiers de l'AARS 10:111–126.*

Henry, Donald O., Harold J. Hietala, Arlene M. Rosen, Yuri E. Demidenko, Vitaliy I. Usik, and Teresa L. Armagan. 2004. Human Behavioral Organization in the Middle Paleolithic: Were Neanderthals Different? *American Anthropologist* 106 (1):17–31.

Herzfeld, Michael. 1992. Metapatterns: Archaeology and the Uses of Evidential Scarcity. In *Representations in Archaeology*, edited by Jean-Claude Gardin and Christopher S. Peebles, pp. 66–86. Bloomington: Indiana University Press.

Heyerdahl, Thor. 1978. *Early Man and the Ocean*. New York: Vintage (Random House).

Hobbes, Thomas. 1651. *Leviathan, or the Matter, Form and Power of a Commonwealth, Ecclesiastical and Civil*. First published in London.

Hodder, Ian. 2006. *The Leopard's Tale: Revealing the Mysteries of Çatalhöyük*. London: Thames and Hudson.

Hodgen, Margaret T. 1974. *Anthropology, History, and Cultural Change*. Tucson: University of Arizona Press.

Hornung, Erik. 2001. *The Secret Lore of Eygpt: Its Impact on the West*. Translated by David Lorton. Ithaca, NY: Cornell University Press. (Original: *Das esoterische Aegypten*, 1999, Munich: C. H. Beck'sche.)

Houston, Stephen D., editor. 2004. *The First Writing: Script Invention as History and Process*. Cambridge: Cambridge University Press.

Hunt, Terry L., and Carl P. Lipo. 2006. Late Colonization of Easter Island. *Science* 311:1603–1606.

Jablonski, Nina G., editor. 2002. *The First Americans: The Pleistocene Colonization of the New World*. San Francisco: California Academy of Sciences.

Jett, Stephen C., 2002. Nicotine and Cocaine in Egyptian Mummies and THC in Peruvian Mummies: A Review of the Evidence and of Scholarly Reaction. *Pre-Columbiana* 2 (4):297–313.

—. 2008. Ancient Ocean Crossings: The Case for Pre-Columbian Contacts Reconsidered. Ms. in author's possession.

Jones, Terry L., and Kathryn A. Klar. 2005. Diffusionism Reconsidered: Linguistic and Archaeological Evidence for Prehistoric Polynesian Contact with Southern California. *American Antiquity* 70 (3):457–484.

Keeley, Lawrence H. 1996. *War Before Civilization: The Myth of the Peaceful Savage*. New York: Oxford University Press.

Kehoe, Alice Beck. 1981. Revisionist Anthropology: Aboriginal North America. *Current Anthropology* 22:503–509, 515–516.

—. 1991a. The Weaver's Wraith. In *The Archaeology of Gender*, edited by Dale Walde and Noreen Willows, pp. 430–435. Calgary: Archaeological Association, University of Calgary.

—. 1991b. No Possible, Probable Shadow of Doubt. *Antiquity* 65 (246):129–131.

—. 1992. Conflict is a Western Worldview. In *The Anthropology of Peace*, edited by Vivian J. Rohrl, M. E. R. Nicholson, and Mario D. Zamora, pp. 55–65. Studies in Third World Societies, No. 47. Williamsburg, VA: College of William and Mary, Department of Anthropology. Reprinted 2000 in *Social Justice: Anthropology, Peace and Human Rights* 1 (1–4):55–61, edited Robert A. Rubinstein, IUAES Commission on Peace & Human Rights. Syracuse, NY: Syracuse University Maxwell School.

—. 1998. *The Land of Prehistory: A Critical History of American Archaeology*. New York: Routledge.

—. 1999. "A Resort to Subtler Contrivances." In *Manifesting Power*, edited by Tracy Sweely, pp. 17–29. London: Routledge.

—. 2000a. François' House, a Significant Pedlars' Post on the Saskatchewan. In *Material Contributions to Ethnohistory: Interpretations of Native North American Life*, edited by Michael S. Nassaney and Eric S. Johnson, pp. 73–187. University Press of Florida, Gainesville, and Society for Historical Archaeology.

—. 2000b. *Shamans and Religion: An Anthropological Exploration in Critical Thinking*. Prospect Heights, IL: Waveland Press.

—. 2000c. Response to O'Brien and Lyman. *Review of Archaeology* 21 (2):33–38.

—. 2003. The Fringe of American Archaeology: Trans-oceanic and Transcontinental Contacts in Prehistoric America. *Journal of Scientific Exploration* 17 (1):19–36.

—. 2004. When Theoretical Models Trump Empirical Validity, Real People Suffer. *Anthropology News* 45 (4):10.

—. 2005. *The Kensington Runestone: Approaching a Research Question Holistically*. Long Grove, IL: Waveland Press.

—. 2006. *North American Indians: A Comprehensive Account*. 3rd edition. Upper Saddle River, NJ: Prentice-Hall.

—. 2007. Osage Texts and Cahokia Data. In *Ancient Objects and Sacred Realms*, edited by F. Kent Reilly III and James Garber. Austin: University of Texas Press,

Kehoe, Alice B., and Thomas F. Kehoe. 1979. *Solstice-Aligned Boulder Configurations in Saskatchewan*. Canadian Ethnology Service Paper No. 48, Mercury Series. Ottawa: National Museum of Man.

Kehoe, Alice Beck, and Thomas C. Pleger. 2007. *Archaeology: A Concise Introduction*. Long Grove, IL: Waveland Press.

Kehoe, Thomas F. 1973. *The Gull Lake Site: A Prehistoric Bison Drive Site in Southwestern Saskatchewan*. Publications in Anthropology and History No. 1. Milwaukee, WI: Milwaukee Public Museum.

—. 1990. Corralling Life. In *The Life of Symbols*, edited by Mary LeCron Foster and Lucy Jayne Boscharow, pp. 175–193. Boulder, CO: Westview Press.

Kelley, David H. 1960. Calendar Animals and Deities. *Southwestern Journal of Anthropology* 16 (3):317–337.

—. 1981. The Invention of the Mesoamerican Calendar. Unpublished paper in possession of author.

Kelley, David H., and Eugene F. Milone. 2005. *Exploring Ancient Skies: An Encyclopedic Survey of Archaeoastronomy*. New York: Springer.

Kelley, Jane H., and Marsha P. Hanen. 1988. *Archaeology and the Methodology of Science*. Albuquerque: University of New Mexico Press.

Kennedy, Roger G. 1994. *Hidden Cities*. New York: Free Press and Penguin.

Kerber, Jordan E., editor 2006. *Cross-Cultural Collaboration: Native Peoples and Archaeology in the Northeastern United States*. Lincoln: University of Nebraska Press.

Kirch, Patrick Vinton, and Roger C. Green. 2001. *Hawaiki, Ancestral Polynesia*. Cambridge: University of Cambridge Press.

Kristiansen, Kristian, and Thomas B. Larsson. 2005. *The Rise of Bronze Age Society: Travels, Transmissions and Transformations*. Cambridge: Cambridge University Press.

Kroeber, Alfred L. 1952. *The Nature of Culture*. Chicago: University of Chicago Press.

Krupp, E. C. 1978. *In Search of Ancient Astronomies*. Garden City, NY: Doubleday.

La Flesche, Francis. 1925. The Osage Tribe: Rite of Vigil. 39th *Annual Report*, Bureau of American Ethnology, Smithsonian Institution, pp. 31–635. Washington, DC: Government Printing Office.

—. 1930. The Osage Tribe: Rite of the Wa-Xo'-Be. 45th *Annual Report*, Bureau of American Ethnology, Smithsonian Institution, pp. 528–833. Washington, DC: Government Printing Office.

Lakoff, George, and Mark Johnson. 1980. *Metaphors We Live By*. Chicago: University of Chicago Press.

Lathrap, Donald W. 1977. Our Father the Cayman, Our Mother the Gourd: Spinden Revisited, or a Unitary Model for the Emergence of Agriculture in the New World. In *Origins of Agriculture*, edited by Charles A Reed, pp. 713–751. The Hague: Mouton.

Latin American Antiquity 17(1), 2006. San Lorenzo researchers (Hector Neff, Jeffrey Blomster, Michael Glascock, Ronald Bishop, James Blackman, Michael Coe, George Cowgill, Richard Diehl, Stephen Houston, Arthur Joyce, Carl Lipo, Barbara Stark, and Marcus Winter): 54–76; Oaxaca researchers (Robert Sharer, Andrew Balkansky, James Burton, Gary Feinman, Kent Flannery, David Grove, Joyce Marcus, Robert Moyle, Douglas Price, Elsa Redmond, Robert Reyolds, Prudence Rice, Charles Spencer, James Stoltman, and Jason Yaeger): 90–103; San Lorenzo researchers (Hector Neff, Jeffrey Blomster, Michael Glascock, Ronald Bishop, James Blackman, Michael Coe, George Cowgill, Stephen Houston, Arthur Joyce, Carl Lipo, Ann Cyphers, and Marcus Winter):104–118.

Lavallée, Danièle. 2000. *The First South Americans*. Translated by Paul G. Bahn. Salt Lake City: University of Utah Press.

Leach, Edmund. 1991. Aryan Invasions over Four Millennia. In *Culture through Time: Anthropological Approaches*, edited by Emiko Ohnuki-Tierney, pp. 227–245. Stanford, CA: Stanford University Press.

LeBlanc, Steven A. 1999. *Prehistoric Warfare in the American Southwest*. Salt Lake City: University of Utah Press.

Lee, Richard B. 1968. Comments. In *New Perspectives in Archeology*, edited by Sally R. Binford and Lewis R. Binford, pp. 343–346. Chicago: Aldine.

Lehner, Mark. 1997. *The Complete Pyramids*. London: Thames and Hudson.

Lepper, Bradley T. 1992. Radiocarbon Dates for Dinosaur Bones? A Critical Look at Recent Creationist Claims. *Creation/Evolution* 12 (1):1–9.

Lesko, Leonard H., editor. 1994. *Pharaoh's Workers: The Villagers of Deir el Medina*. Ithaca: Cornell University Press.

Lightfoot, Kent G. 2005. *Indians, Missionaries, and Merchants: The Legacy of Colonial Encounters on the California Frontiers*. Berkeley: University of California Press.

Lyons, Patrick D. 2003. *Ancestral Hopi Migrations*. Tucson: University of Arizona Press.

Mason, Ronald J. 2006. *Inconstant Companions*. Tuscaloosa: University of Alabama Press.

Mayor, Adrienne. 2005. *Fossil Legends of the First Americans*. Princeton: Princeton University Press.

Mayr, Ernst. 1982. *The Growth of Biological Thought: Diversity, Evolution, and Inheritance*. Cambridge, MA: Belknap.

McGuire, Randall H. 1991. *A Marxist Archaeology*. San Diego, CA: Academic Press.

McIntosh, Roderick J. 2005. *Ancient Middle Niger: Urbanism and the Self-Organizing Landscape*. Cambridge: Cambridge University Press.

Merrien. Jean [real name, René Marie de la Poix de Fréminville]. 1954. *Lonely Voyagers [Les Navigateurs Solitaires]*, English translation 1954, by J. H. Watkins. New York: G. P. Putnam's Sons.

Michino, Gregory F., 1997. *Lakota Noon: The Indian Narrative of Custer's Defeat*. Missoula, MT: Mountain Press.

Morgan, Lewis Henry. 1985 [1877]. *Ancient Society*. Tucson: University of Arizona Press. Facsimile.

Moser, Stephanie. 2006. *Wondrous Curiosities: Ancient Egypt at the British Museum*. Chicago: University of Chicago Press.

Needham, Joseph. 1986. *Science and Civilisation in China*, Vol. 5, pt. 7 (Military Technology; The Gunpowder Epic). Cambridge: Cambridge University Press.

Needham, Joseph, and Gwei-Djen Lu. 1985. *Trans-Pacific Echoes and Resonances; Listening Once Again*. Singapore: World Scientific.

Nichols, Johanna. 1992. *Linguistic Diversity in Space and Time*. Chicago: University of Chicago Press.

Nielsen, Richard, and Scott F. Wolter. 2006. *The Kensington Runestone: Compelling New Evidence*. St. Paul, MN: Lake Superior Agate Publishing.

Novick, Peter. 1988. *That Noble Dream: The "Objectivity Question" and the American Historical Profession*. Cambridge: Cambridge University Press.

Numbers, Ronald L. 1992. *The Creationists*. NewYork: Knopf.

O'Brien, Michael J., R. Lee Lyman, and Michael Brian Schiffer. 2005. *Archaeology as a Process*. Salt Lake City: University of Utah Press.

Onuf, Peter S. 1997. Thomas Jefferson, Missouri, and the '"Empire for Liberty." In *Thomas Jefferson and the Changing West: From Conquest to Conservation*, edited by James P. Ronda, pp. 111–153. St. Louis: Missouri Historical Society Press.

Overstreet, David F. 1998. Late Pleistocene Geochronology and the Paleoindian Penetration of the Southwestern Lake Michigan Basin, *Wisconsin Archeologist* 79 (1):28–52.

Parezo, Nancy J., editor. 1994. *Hidden Scholars: Women Anthropologists and the Native American Southwest*. Albuquerque: University of New Mexico Press.

Pauketat, Timothy R. 2004. *Ancient Cahokia and the Mississippians*. Cambridge: Cambridge University Press.

—. 2007. *Chiefdoms and Other Archaeological Delusions*. Lanham, MD: AltaMira Press (Rowan and Littlefield).

—. 2008. *Cahokia's Big Bang and the Story of Ancient North America*. New York: Viking/ Penguin.

Peel, John D. Y. 1971. *Herbert Spencer: The Evolution of a Sociologist*. London: Heinemann.

Pettitt, Paul. 2005. The Rise of Modern Humans. In *The Human Past: World Prehistory and the Development of Human Societies*, edited by Chris Scarre, pp. 124–173. London: Thames and Hudson.

Petto, Andrew J., and Laurie R. Godfrey. 2007. *Scientists Confront Creationism and Intelligent Design*. New York: W. W. Norton.

Phillips, David A., Jr., and Lynne Sebastian, editors. 2001. *Exploring the Course of Southwest Archaeology: The Durango Conference, September 1995*. Albuquerque: New Mexico Archaeological Council.

Pleger, Thomas C. 2000. "Old Copper and Red Ocher Social Complexity," *Midcontinental Journal of Archaeology* 25 (2):169–190.

Powell, John Wesley. 1881. Report of the Director. First *Annual Report*, Bureau of Ethnology, 1879—80, pp. xi–xxxiii. Washington: Government Printing Office.

Preston, Douglas. 1995. The Mystery of Sandia Cave. *The New Yorker* 71 (June 12, 1995): 66–83.

—. 1999. Woody's Dream. *The New Yorker* 75 (34):80–87.

Prufer, Olaf H. 2001. The Archaic of Northeastern Ohio. In *Archaic Transitions in Ohio and Kentucky Prehistory*, edited by Olaf H. Prufer, Sara E. Pedde, and Richard S. Meindl, pp. 183–209. Kent, OH: Kent State University Press.

Quinn, David B. 1974. *England and the Discovery of America, 1481–1620*. London: Allen and Unwin.

Ridington, Robin, and Dennis Hastings (In'aska). 1997. *Blessing for a Long Time: The Sacred Pole of the Omaha Tribe* . Lincoln: University of Nebraska Press.

Rogers, Everett M. 1962. *Diffusion of Innovations*. New York: Free Press.

Scarre, Chris, editor. 2005. *The Human Past: World Prehistory and the Development of Human Societies*. London: Thames and Hudson.

Schmandt-Besserat, Denise. 1992. *Before Writing: From Counting to Cuneiform*. Austin: University of Texas Press.

Schwartz, Stephan A. 1978. *The Secret Vaults of Time: Psychic Archaeology and the Quest for Man's Beginnings*. New York: Grosset and Dunlap.

Seifert, Donna J. 1991. Within Site of the White House: The Archaeology of Working Women. *Historical Archaeology* 25 (4):82–108.

Severin, Tim. 1978. *The Brendan Voyage*. New York: Avon.

Shao, Paul. 1976. *Asiatic Influences in Pre-Columbian American Art*. Ames: Iowa State University Press.

Shapin, Steven, and Simon Schaffer 1985. *Leviathan and the Air-Pump: Hobbes, Boyle and the Experimental Life*. Princeton: Princeton University Press.

Sidky, Houmaym. 2007. Haunted by the Archaic Shaman: Himalayan Jhākris and the Discourse on Shamanism. Ms. under submission to publisher.

Silliman, Stephen W. 2004. *Lost Laborers in Colonial California: Native Americans and the Archaeology of Rancho Petaluma*. Tucson: University of Arizona Press.

Silverman, Helaine, editor. 2006. *Archaeological Site Museums in Latin America*. Gainesville: University Press of Florida.

Smith, Adam T. 2003. *The Political Landscape: Constellations of Authority in Early Complex Polities*. Berkeley: University of California Press.

Sorenson, John L. 1998. *Images of Ancient America: Visualizing Book of Mormon Life*. Provo, UT: Research Press of the Foundation for Ancient Research and Mormon Studies

Sorenson, John L., and Carl L. Johannessen, 2004. *Scientific Evidence for Pre Columbian Transoceanic Voyages to and from America*. Sino Platonic Papers 133, CD-ROM edition. Philadelphia: University of Pennsylvania, Department of Asian and Middle Eastern Studies.

—. 2006. Biological Evidence for Pre-Columbian Transoceanic Voyages. In *Contact and Exchange in the Ancient World*, edited by Victor H. Mair, pp. 238–297. Honolulu: University of Hawai'i Press.

Stafford, Thomas W. 1992. Radiocarbon Dating Dinosaur Bones: More Pseudoscience from Creationists. *Creation/Evolution* 12 (1):10–17.

Stevenson, Christopher M., Joan Wozniak, and Sonia Haoa. 1999. Prehistoric Agricultural Production on Easter Island (Rapa Nui), Chile. *Antiquity* 73 (282):801–812.

Steward, Julian H. 1955. *Theory of Culture Change*. Urbana: University of Illinois Press.

Storey, Alice A., Jose Miguel Ramirez, Daniel Quiroz, David V. Burley, David J. Addison, Richard Walter, Atholl J. Anderson, Terry L. Hunt, J. Stephen Athens, Leon Huynens, and Elizabeth Matisoo-Smith. 2007. Radiocarbon and DNA Evidence for Pre-Columbian Introduction of Polynesian Chickens to Chile. *Proceedings of the National Academy of Sciences* 104 (25):10335–10339.

Straus, Lawrence Guy. 2000. Solutrean Settlement of North America? A Review of Reality. *American Antiquity* 65 (2):219–226 .

Sundstrom, Linéa. 2004. *Storied Stone: Indian Rock Art of the Black Hills Country*. Norman: University of Oklahoma Press.

Thomas, David Hurst. 2000. *Skull Wars: Kennewick Man, Archaeology, and the Battle for Native American Identity*. New York: Basic Books.

Thompson, Thomas L. 1999. *The Mythic Past: Biblical Archaeology and the Myth of Israel*. New York: Basic Books.

Trigger, Bruce A. 1998. *Sociocultural Evolution*. Oxford: Blackwell.

Tubb, Kathryn Walker. 2006. Artifacts and Emotion. In *Archaeology, Cultural Heritage, and the Antiquities Trade*, edited by N. Brodie, M. M. Kersel, C. Luke, and K. W. Tubb, pp. 284–302. Gainesville: University Press of Florida.

Van Tilburg, Jo Anne. 1994. *Easter Island*. Washington, DC: Smithsonian Institution Press.

Walde, Dale A. 2006. Avonlea and Athabaskan Migrations: A Reconsideration. *Plains Anthropologist* 51 (198):185–197.

Wall, Diana diZerega, and Anne-Marie Cantwell. 2004. *Touring Gotham's Archaeological Past*. New Haven: Yale University Press.

Wallace, Anthony F. C. 1962. *Culture and Personality*. New York: Random House. Reprinted, p. 213, in *Revitalizations and Mazeways*, by Anthony F. C. Wallace and edited by Robert S. Grumet, 2003, Lincoln: University of Nebraska Press.

—. 1997. "The Obtaining Lands": Thomas Jefferson and the Native Americans. In *Thomas Jefferson and the Changing West: From Conquest to Conservation*, edited by James P. Ronda, pp. 25–41. St. Louis: Missouri Historical Society Press.

—. 1999. *Jefferson and the Indians*. Cambridge, MA: Belknap.

Watkins, Joe. 2000. *Indigenous Archaeology: American Indian Values and Scientific Practice*. Walnut Creek, CA: AltaMira.

Webb, S. [Steve] C. 2006. *The First Boat People*. [Pleistocene Australians.] Cambridge: Cambridge University Press.

Wilcox, David R., and Jonathan Haas. 1994. The Scream of the Butterfly: Competition and Conflict in the Prehistoric Southwest. In *Themes in Southwest Prehistory*, edited by George J. Gumerman, pp. 211–238. Santa Fe, NM: School of American Research Press.

Wilkins, David E., and K. Tsianina Lomawaima. 2001. *Uneven Ground: American Indian Sovereignty and Federal Law*. Norman: University of Oklahoma Press.

Wood, Margaret C. 2002. Women's Work and Class Conflict in a Working-Class Coal-Mining Community. In *The Dynamics of Power*, edited by Maria O'Donovan, pp. 66–87. Occasional Paper No. 30. Carbondale: Center for Archaeological Investigations, Southern Illinois University.

World Trade Organization Global Code of Ethics for Tourism: http://www.world-tourism.org/code_ethics/eng.html

Yoffee, Norman. 2005. *Myths of the Archaic State: Evolution of the Earliest Cities, States, and Civilizations*. Cambridge: Cambridge University Press.

INDEX

About the Author

Alice Beck Kehoe is Professor of Anthropology, emerita, Marquette University, and Adjunct Professor, Anthropology, University of Wisconsin-Milwaukee. She has been President of the Central States Anthropological Society, served on the Board of Directors, American Anthropological Association, and was chair of the Society for American Archaeology's Public Education Committee. Among her publications are the widely used textbook, *North American Indians: A Comprehensive Account* (Prentice-Hall, third edition, 2006); *Land of Prehistory: A Critical History of American Archaeology* (Routledge, 1998); *America before the European Invasions* (Longman, 2002); and *Archaeology: A Concise Introduction* (Waveland, 2007, co-authored with Thomas Pleger).

Kehoe excavated bison drives, a fur trade post, and an archaeoastromical site in the Northwestern Plains. She conducted ethnoarchaeological research in northern Saskatchewan and with Aymara on Lake Titicaca, Bolivia, and participated in excavations at Solutré in France, Dolní Vestonice in the Czech Republic, Tiwanaku in Bolivia, and as a student in northwest Mexico, Indiana, and southern Illinois. Her current academic interests include the history of archaeology, pre-contact, ethnohistorical, and contemporary research on American First Nations, and continued collaboration with the Tribal Historic Preservation Office, Piegan Institute, and Blackfeet Community College on the Blackfeet Reservation.